Chemists' Guide to Effective Teaching

Chemists' Guide to Effective Teaching

Edited by

Norbert J. Pienta
Department of Chemistry
University of Iowa

Melanie M. Cooper
Department of Chemistry
Clemson University

Thomas J. Greenbowe
Department of Chemistry
Iowa State University

Prentice Hall Series in Educational Innovation

PEARSON
Prentice Hall

Upper Saddle River, NJ 07458

Library of Congress Cataloging-in Publication Data

Chemists' guide to effective teaching / [edited by] Norbert J. Pienta, Melanie M. Cooper, Thomas J. Greenbowe.
    p. cm.
Includes bibliographical references.
ISBN 0-13-149392-2 (pbk.)
 1. Chemistry--Study and teaching (Higher) I. Pienta, Norbert J. II. Cooper, Melanie M. III. Greenbowe, Thomas J.

QD40.C454 2005
540'.71'1--dc22

                                           2004028660

Executive Editor: *Kent Porter Hamann*
Project Manager: *Jacquelyn Howard*
Editor-in-Chief, Science: *John Challice*
Production Editor: *Shari Toron*
Assistant Managing Editor: *Beth Sweeten*
Manufacturing Buyer: *Alan Fischer*
Marketing Manager: *Steve Sartori*
Art Director: *Jayne Conte*
Cover Designer: *Bruce Kenselaar*
Cover Illustration: *Getty Images, Inc.*

 © 2005 Pearson Education, Inc.
Pearson Prentice Hall
Pearson Education, Inc.
Upper Saddle River, New Jersey 07458

Printed in the United States of America
10 9 8 7 6 5 4 3 2 1

ISBN 0-13-149392-2

Pearson Education LTD., *London*
Pearson Education Australia PTY., Limited, *Sydney*
Pearson Education Singapore, Pte. Ltd.
Pearson Education North Asia Ltd., *Hong Kong*
Pearson Education Canada, Ltd., *Toronto*
Pearson Educación de Mexico, S.A. de C.V.
Pearson Education—Japan, *Tokyo*
Pearson Education Malaysia, Pte. Ltd.

Chemists' Guide to Effective Teaching

Edited by

Norbert J. Pienta
Department of Chemistry
University of Iowa

Melanie M. Cooper
Department of Chemistry
Clemson University

Thomas J. Greenbowe
Department of Chemistry
Iowa State University

Prentice Hall Series in Educational Innovation

PEARSON
Prentice
Hall

Upper Saddle River, NJ 07458

Titles in the Prentice Hall Series in Educational Innovation

Chemistry ConcepTests: A Pathway to Interactive Classrooms
By Clark R. Landis, Arthur B. Ellis, George C. Lisenky, Julie K. Lorenz, Kathleen Meeker, and Carl C. Wasmer
0-13-090628-X

Chemists' Guide to Effective Teaching
Edited by Norbert J. Pienta, Melanie M. Cooper, and Thomas J. Greenbowe
0-13-149392-2

The Ethical Chemist
By Jeffrey D. Kovac
0-13-141132-2

Introductory Chemistry: A Workbook
By Robert E. Blake, Jr.
0-13-144602-9

Peer-Led Team Learning: A Guidebook
By David K. Gosser, Mark S. Cracolice, J.A. Kampmeier, Vicki Roth, Victor S. Strozak, and Pratibha Varma-Nelson
0-13-028805-5

Peer-Led Team Learning: A Handbook for Team Leaders
By Vicki Roth, Ellen Goldstein, and Gretchen Mancus
0-13-040811-5

Peer-Led Team Learning: General Chemistry
By David K. Gosser, Victor S. Strozak, and Mark S. Cracolice
0-13-028806-3

Peer-Led Team Learning: General, Organic, and Biological Chemistry
By Pratibha Varma-Nelson and Mark S. Cracolice
0-13-028361-4

Peer-Led Team Learning: Organic Chemistry
By J.A. Kampmeier, Pratibha Varma-Nelson, and Donald Wedegaertner
0-13-028413-0

Science and Its Ways of Knowing
By John Hatton and Paul B. Plouffe
0-13-205576-7

Survival Handbook for the New Chemistry Instructor
Edited by Diane M. Bunce and Cinzia M. Muzzi
0-13-143370-9

Writing Across the Chemistry Curriculum: An Instructor's Handbook
By Jeffrey Kovac and Donna Sherwood
0-13-029284-2

CONTENTS

Prologue
Norbert J. Pienta, Melanie M. Cooper, and Thomas J. Greenbowe..ix

PART I: COGNITION

Chapter 1: Introduction to Chemists' Guide to Effective Teaching...2
J. Dudley Herron

Chapter 2: How Students Learn: Knowledge Construction in College Chemistry Courses.................12
Mark S. Cracolice

Chapter 3: All Students Are Not Created Equal: Learning Styles in the Chemistry Classroom...........28
Stacey Lowery Bretz, Ph.D.

Chapter 4: Inquiry and the Learning Cycle Approach...41
Michael R. Abraham

Chapter 5: Relevance and Learning Theories...53
Donald J. Wink

Chapter 6: Models and Modeling...67
George M. Bodner, David E. Gardner, and Michael W. Briggs

Chapter 7: Enhancing Students' Conceptual Understanding of Chemistry through Integrating the Macroscopic, Particle, and Symbolic Representations of Matter.......................................77
Dorothy Gabel

PART II: TEACHING STRATEGIES

Chapter 8: The Role of Analogies in Chemistry Teaching...90
MaryKay Orgill and George M. Bodner

Chapter 9: Solving Word Problems in Chemistry: Why Do Students Have Difficulty and What Can Be Done to Help?...106
Diane M. Bunce

Chapter 10: An Introduction to Small-Group Learning...117
Melanie M. Cooper

Chapter 11: Using Concept Maps to Figure Out What Your Students Are Really Learning..............129
Mary B. Nakhleh and Yilmaz Saglam

Chapter 12: Introduction to the Science Writing Heuristic...140
Thomas J. Greenbowe and Brian Hand

Chapter 13: Team Learning...155
Pratibha Varma-Nelson and Brian P. Coppola

PART III: LEARNING WITH TECHNOLOGY

Chapter 14: Electronic Data Collection to Promote Effective Learning During Laboratory Activities..172
I. Dwaine Eubanks

Chapter 15: Wireless Inside and Outside of the Classroom..186
Jimmy Reeves and Charles R. Ward

Chapter 16: Using Multimedia to Visulaize the Molecular World: Educational Theory into Practice..195
Roy Tasker

Prologue

Our intention for this book was to provide chemists and scientists, who are assigned to teach a course about chemistry, an introduction to some relevant aspects of learning theory that apply to teaching chemistry. We hope that those who read it will see one or two techniques that they are willing to try in their classroom. Each chapter is written by a chemist who has some expertise in the topic, has done some research on the topic, and has applied the technique discussed in their chapter in their chemistry courses. This book does not include everything one should know about learning theory. There are some major topics, such as information processing theory, that are not included because we wanted to focus on topics for which chemists can see an immediate pay-off with their students.

Who should use this book?

One of us recently visited a research university to present a seminar in the department of chemistry. During that visit, a colleague quipped, "I know how to be a good teacher...teaching is not rocket science..." He was certainly serious, and perhaps he even did know what it meant to be a "good teacher." As a good researcher, he would certainly read the literature and keep up with the latest findings and techniques. So what about teaching? Why not expect the same for ourselves and our colleagues when it comes to teaching chemistry to undergraduate students? Most universities and colleges have a renewed interest in doing a better job in educating undergraduate students. The National Science Foundation now requires researchers to include an educational component or outreach component in scientific grant proposals. This book is directed at chemists who teach—any and all of them. It is not enough for an instructor to present a clear, well-organized lecture. Instructors must now demonstrate that their students are actively involved in the learning.

Everyone can and should expect to get better at what they do, and teaching is no exception. In fact, given the objective nature of most scientific or chemical research data, many academic chemists may well be very suspicious and unconvinced when looking at chemical education data that is based on human behavior. But the data really are there, and experiments that are well done are quite convincing. Part of the problem is to recognize good science education and/or chemical education research. We admit that there are published science education research articles for which data are incomplete or at best inconclusive. A future volume will include an analysis of the 10–15 best chemical education research experiments that have had an impact on how we teach chemistry or how students learn chemistry. For example, there have been several excellent studies on how students can do algorithmic-type problems (i.e., they can generate a number by using a formula), but fail to demonstrate conceptual understanding of that same topic. (By the way, we encourage you to include one or two conceptual questions on your next exam and see the results.) Let the ideas in the following chapters be the basis for improving your teaching. Then give each of your colleagues a copy. And it's not because the editors might share a bunch of royalties for this book; we're not accepting any, and besides, we're serious about this. However, we would be happy to see this book and the contents hotly debated by chemists rather than gaining fame and fortune for us.

Few graduate students in chemistry receive any formal or informal instruction about educational theories, pedagogy, and successful strategies for helping students learn. Graduate students and new faculty often have some things in common when it comes to teaching. They are enthusiastic, often zealots about the content. They also have little or no teaching experience. And without a theoretical underpinning and exposure to different, successful pedagogies, they will revert to that which is familiar (i.e., rote lecture). By the time we reach the professorial stage, we are very experienced at lecturing and listening to presentations involving the lecture style. Many experienced chemistry instructors know how to receive high marks from students on "student evaluations of the lecture." This does not mean that these instructors necessarily are effective teachers or that the students learned the subject matter. In part, this is because the novice learner knows neither the basics of the content nor how he or she learns things. Instruction accounts for the remainder of the problem. This book provides that theory and certainly shows that there are alternatives to lecture and ways to encourage more active student participation. One might argue that TAs and new faculty might be better served by reading a similar book that is organized as a primer or one that concentrates on practical applications, and we are happy to suggest the first volume of this series, "Survival Handbook for the New Chemistry Instructor" (Bunce and

Muzzi, eds, Prentice Hall, 2004). But there are plenty of examples and suggestions amid the explanations in this volume for that group.

So what can a seasoned veteran learn about teaching? After several years, most instructors carefully and successfully make the observations that some things don't work and others are highly successful. You can find out why and, if you are industrious, you too can design a study to collect data to demonstrate your point. If you are out of new ideas, each chapter certainly offers plenty of ideas to try, many of which have a sound pedagogical basis that is supported by data. College chemistry courses, particularly introductory ones, enroll students with diverse abilities, interests, and levels of motivation. The components of the course should also be multi-faceted in helping that diverse group learn. In addition, if you are a real cynic, you will find yourself better prepared to debate your colleagues in education and to better understand why some colleagues in your department want to hire one or more chemical educators.

Finally, if you are a chemical educator, we hope this book serves as reference for a graduate course in chemical education, to inspire a new research area, or to publicly (or secretly) present to some of your colleagues.

How should one use this book?

It would be challenging to read this book like a novel, although we did organize it in a rational way with individual chapters able to stand on their own. Some chapters may make more sense after you read others or after you finish a whole group. The editors encouraged the authors to give examples from chemistry, and when they did, we asked for more. Each chapter represents one or more things that you can try. However, the practical applications and the "how to" instructions will appear in another volume. Although the chapters are not organized in this way, there is something for instructors at all levels—teaching assistant training materials, theory and strategies for novice professors who did not have substantive instruction on learning theory for teaching chemistry, and an opportunity to keep up with the changing times for the more senior and experienced colleagues.

Oh yes, there is more…

Since this book is a monograph on learning theory, it serves as a primer or introduction to each topic. We asked each author to suggest relevant articles for you to read on each topic. The suggested articles provide more detail, and most include chemistry examples or context. These citations and the references should allow interested readers to delve into a subject. May we also suggest that you take the time to voice your opinion about teaching and learning when you have the opportunity to talk with your colleagues. When attending meetings of the American Chemical Society, engage the speakers and participants who are participating in a chemical education symposium in a discussion about teaching and learning. In your classroom, try something other than lecturing to your students. You may find that your students learn the topic better, retain the ideas of the topic longer, and enjoy learning.

Acknowledgments

The editors would like to thank Kent Porter Hamann, Jackie Howard, and the management at Prentice Hall for making this happen. Tom and then Melanie and Norb separately approached John Challice and Kent with the idea. We combined forces and ideas and asked our chemical education colleagues for help. We acknowledge the hard work and knowledge of our very capable team of authors. This is not the extent of chemical education knowledge or research. Our quest continues, we hope to include many other authors in a future volume, and we hope to earn your continuing interest.

Norbert J. Pienta
Melanie M. Cooper
Thomas J. Greenbowe

Part

I

COGNITION

Introduction to Chemists' Guide to Effective Teaching

J. Dudley Herron, Chair (retired)
Department of Physical Sciences
Morehead State University, KY

Abstract

This chapter presents a personal account of how chemical education evolved at Purdue University, followed by comments about each chapter of this book. The Department of Chemistry at Purdue University was one of the first in the nation to hire a chemical educator and one of the first to grant chemical education equal status with inorganic, analytical, biochemical, organic, physical, and other divisions in the department. Today, there are at least 23 chemistry departments in the USA that have hired chemical educators and at least 16 have a Ph.D. program in chemical education (Bretz, 2004; Mason, 2001). Such appointments facilitate interactions between chemical education researchers and other research chemists; these interactions must occur in order to improve students' understanding of chemistry.

This book need not be read in chapter order. Brief comments about each chapter enable readers to select materials of greatest interest and gain insight into how each chapter relates to ideas presented elsewhere in the book.

Biography

Before retiring from Morehead State University in 1996, Professor Herron taught high school chemistry for four years, taught chemistry and education at Purdue University for 25 years, and spent four years in administration—two years as Chair of the Department of Curriculum and Instruction at Purdue University and two and one-half years as Chair of the Department of Physical Sciences at Morehead State University.

During his university career, Professor Herron was active in research and curriculum development at many levels. He was on the author team that developed the *Intermediate Science Curriculum Study* (ISCS), an individualized, laboratory-centered program for middle school science, and he coordinated the field trial of those materials in Indiana. He was senior author of *Heath Chemistry* (2nd ed., 1993) and the author of *Understanding Chemistry: A Preparatory Course* (2nd ed., 1986) and *The Chemistry Classroom: Formulas for Successful Teaching* (ACS Books, 1996), a book summarizing much of his research in chemistry education. He has been a frequent contributor to the *Journal of Chemical Education, Journal of Research in Science Teaching,* and similar professional journals. Professor Herron's awards include Visiting Scientist of the Year (Western Connecticut Section of ACS, 1982); Catalyst of the Year (Chemical Manufacturers Association, 1983); Outstanding Science Educator (Association for the Education of Teachers of Science, 1985); and the Lilly Endowment Faculty Open Fellowship (1982).

Introduction

If you are a chemist hired as a faculty member in a college or university, you are teaching a chemistry course. Traditional lectures work for about the top 15% of students in most courses. If you are teaching an honors course or at a highly selective school such as Massachusetts Institute of Technology, California Institute of Technology, or the University of California at Berkeley, that percentage will be somewhat higher, but even then it will not approach 100%. What are you going to do for the other 85% of the students in your course? You would not go into a chemistry laboratory and try to synthesize a compound or try to separate a complex mixture without knowing some of the chemistry theory and practical techniques that others have published in the area. This book's editors set out to provide the same kind of background to support your efforts in the chemistry classroom. In the past forty years, an extended body of peer-reviewed literature about teaching and learning chemistry has developed—far more than most chemists have time to glean from primary sources. This book is written for teaching chemists who want a primer on learning theory and teaching techniques. A third volume (forthcoming) in this series will deal with practical applications.

How Chemical Education has Evolved: One Person's View

In 2004, the field of chemistry education is coming of age. When I began teaching in 1958, chemists and educationists lived in separate worlds, ignoring one another. Self-respecting chemists dismissed educationists as well-meaning simpletons whose "research" amounted to little more than meaningless surveys concerning inconsequential questions that were reported in incomprehensible jargon. Similarly, educationists saw chemists as arrogant prima donnas too obsessed with advancing their research careers to consider how teaching and learning might be improved. At the University of Kentucky, where I did my undergraduate work, Limestone Street separated the College of Education and the College of Arts and Science. That gulf was more difficult to cross than the palisades of the Kentucky River, just a few miles west and south!

There were, of course, exceptions. The curriculum development projects that began in the late 1950s and continued through the early 1970s were characterized by cooperation between scientists, mathematicians, psychologists, and educationists. Those projects fostered respect–even admiration–among participating parties, but such interaction was not widespread. When a project ended, everyone returned to his or her respective niche, and constructive interaction ceased.

Although I began my undergraduate program as a chemistry major, working at a mental hospital to pay for my schooling convinced me that I wanted to work with people more than I wanted to work with chemicals. Robert Wagner, my undergraduate advisor, suggested that I consider teaching, and I never looked back. Unlike many who consider education courses to be a waste of time, I found mine interesting and worthwhile.

In 1958, there were, to my knowledge, no faculty in departments of chemistry doing research on teaching and learning.[1] Science education–there was no chemical education, physics education, and such at the time–was housed in schools of education. Faculty got their positions because they were good pre-college teachers, and their primary responsibility was to pass along their teaching skills to neophytes. Many, like their chemistry colleagues, were pessimistic that research on teaching and learning made any sense, and those who thought it worthwhile were poorly trained to do it. Education had no research laboratories where graduate students served apprenticeships under the careful supervision of a faculty mentor. Nor was there money to support research in science education. It was all going toward curriculum development.

Through John Mayor, then Education Director at AAAS, I was invited to participate in several conferences and workshops related to these "new curricula." I became convinced that significant improvement in science teaching and learning could not take place without the expertise of scientists, who understood what needed to be learned, and educationists, who were familiar with cognitive science and adolescent personality. Without the expertise of both groups, little could be accomplished, and the problems were too complex to be solved quickly. The curriculum development projects provided the right mix–scientists, cognitive psychologists, and teachers– but not the right time frame. Chemists like George Pimentel were willing to take a year or two away from research to "fix" secondary school science, but it wasn't their life's work. At about this time, I began to formulate a professional goal: "Before I die, I would like to see somewhere on the face of Earth an institutional structure where individuals on the cutting edge of research on teaching and learning and individuals on the

---

[1] Hubert Alyea, Derek Davenport, Jay Young, and others were producing excellent laboratory and demonstration materials as well as new techniques for presenting them, but studies of chemistry learning grounded in sound theory were virtually nonexistent.

cutting edge of research in chemistry work together and talk to one another." Purdue University seemed to have potential to be that place.

Purdue had no School of Education when I went there in 1965. Prospective high school teachers majored in the discipline they intended to teach, taking education courses required for certification in the Department of Education. At that time, Purdue's Department of Education was in the School of Humanities, Social Science, and Education. The departments of physics, biology, and mathematics had hired faculty members with high school teaching experience to head up their teacher education programs; chemistry had not. In the spring of 1965, the Chemistry Department at Purdue University advertised such a position, and I applied. The chemists in charge of the search were as dubious of my degree in science education as I was suspicious of a search for a science educator housed in a research-oriented chemistry department. What did I know about chemistry, and why hadn't I gotten a real degree rather than messing around in education? What did they know about teaching, and why weren't faculty in education conducting this search?

I had no qualms about teaching a methods course for chemistry teachers, supervising their student teaching, and the like, but I wasn't confident about teaching college chemistry. I had taught high school chemistry well, and John Mayor had immersed me in the community of scientists, educators, and psychologists who, for the past decade, had been revising the way that science and mathematics were taught.

As a graduate student at Florida State, I had participated in curriculum reform in junior high science. More than most new Ph.D.'s, I had thought about elementary chemistry concepts and principles—how one idea led to another, what chemical systems might be used successfully to illustrate those ideas, and where students typically lost their way and gave up in frustration. Still, the graduate courses in thermodynamics, quantum mechanics, and analytical chemistry that had been part of my master's and doctoral programs left me more aware of my ignorance than confident in my ability to teach chemistry at the college level. I wasn't a chemist, and I knew it. Did they?

Job interviews in Schools of Education—at least those conducted in 1965—focus on what the interviewee knows about teaching, his or her practical experience in the classroom, familiarity with the politics of schools, and teacher certification. Job interviews in chemistry cut to the chase: "O.K., Dudley, tell us about your research." Although I always cringed when my graduate students asked to read my doctoral dissertation, the research it reported was probably above average for science education at the time. But it wasn't what chemists do! Grant Urey, then head of the inorganic division at Purdue, asked the first question: "You call that research?" I knew that my examination of differences in what high school students learned in *Chem Study*, *CBA*, and traditional chemistry courses was about as close to chemistry research as the smell of decaying flesh is to Chanel No. 5!"I know it isn't what you call research. Call it a study, a survey, or whatever you like, but know that this is the kind of question that interests me, and it is the kind of work I expect to do. If you don't want somebody in your department doing this sort of thing, you shouldn't hire me," I replied. The question hadn't bothered me, and I hoped that my answer wouldn't bother them. Apparently it didn't.

Once I was excused from the room, Henry Feuer suggested that I be hired. "Anyone who can stand up to Grant Urey like that," Henry opined, "should do very well here." Henry's tongue-in-cheek observation probably got me the job.

When I went to Purdue, I had the choice of a cramped office in the main chemistry building or more expansive digs in the "temporary" Quonset huts housing freshman chemistry labs near the edge of campus.[1] Fortunately, I selected the remote setting, sheltered from all but fleeting glimpses from other faculty as I adjusted to my new environment and learned enough about the culture of chemistry to survive.

Aware of my weakness, I set about learning enough chemistry to feel confident and not make a fool of myself. I sat in on lectures given by other chemistry faculty, and I was surprised when I heard other new faculty—real chemists—make statements that I knew to be wrong. I had, after all, worked through many common

---

[1] Footnote 2: Nothing survives like a "temporary" building. These FWA buildings at Purdue University were finally demolished in 2004.

misconceptions during my involvement in curriculum development and while sorting out answers to questions asked by high school students. It came as a shock to learn that having a Ph.D. in chemistry didn't ensure that one had achieved total understanding!

Because my Ph.D. was not in chemistry, I readily acknowledged my vast store of ignorance on the subject. When my understanding of an idea was sparse, I didn't hesitate to ask other faculty members about it. It was instructive to learn that they sometimes understood no better than I, and when I went to someone whose research was in the troublesome area, I occasionally found that the answer wasn't totally clear to anyone!

Derek Davenport facilitated my post-graduate education by asking me to check out some general chemistry experiments that weren't working as designed. My sleuthing revealed that assumptions made in adapting an American Standards for Testing and Materials procedure (ASTM International) used to determine the thickness of zinc on galvanized steel had been naive. Furthermore, modifications made in the general chemistry preparation laboratory to reduce the time it took to make the solutions interfered with the chemistry as well. I became keenly aware of what I had long suspected: There is no such thing as a simple chemical reaction!

My messing about in chemistry led to a few improved experiments and assured some faculty that Michael Kasha had been right when he responded to their question about whether I was really a chemist: "No, he isn't a chemist, but he could be." It left me feeling that I might belong in a chemistry department after all.

Although I was beginning to feel that I might belong, I still had much to learn. A year or two after I arrived at Purdue, a new chemistry graduate student asked me to serve as his research director. Honored, though I was, I had to point out that I did not have doctoral directive status. If he wanted to do research in science education, he should talk to Joe Novak in Biology. Within 30 minutes, I received a message that Earl McBee, then Chemistry Department Chair, wanted to see me immediately. He met me outside his office and demanded to know why I was sending a chemistry graduate student to someone in Biology! "But I don't have doctoral directive status," I explained. "Like hell you don't! I don't know about Education, but we don't hire faculty in chemistry who can't direct doctoral research!"

It took years to sort out how this marriage between chemistry, education, and cognitive science might be made to work. The Department of Chemistry had offered a non-thesis master's degree for high school teachers before I arrived, but it had given little consideration to a research degree in chemical education. The research that interested me did not fit the chemistry mold, and an advisory committee composed entirely of chemists would not have the expertise needed to guide my graduate students–not to mention the potential for the student's speedy demise! A typical committee had members from chemistry, statistics, educational research or educational psychology, and science or mathematics education. Master's degrees were available in either the Department of Chemistry or the Department of Education; doctoral degrees were available only in Education.

My first doctoral students were graduates of Purdue's NSF Summer Institute program. They were experienced high school teachers, so they had pre-college teaching experience that was required for admission to graduate study in Education. But what about students who were admitted to graduate study in Chemistry, became interested in teaching as a result of their experience as a teaching assistant, and wanted to do research in chemical education rather than analytical, inorganic, organic, or physical chemistry? Without pre-college teaching experience, they could not be admitted to graduate study in education, and the educational research that they wanted to do was quite foreign to chemistry faculty!

Our programs evolved. In 1983, the Department of Chemistry formed a Division of Chemical Education parallel to other divisions in the department. The department approved a thesis option for chemical education at the master's level, and the Department of Education approved admission to its doctoral program for students whose only teaching experience was at the college level. But shouldn't there be a doctoral program for chemical educators in the Department of Chemistry? Many students identified with chemistry far more than with education, and their career goal was to teach college chemistry and do research in chemical education. Why should their Ph.D. be in Education? George Bodner argued for a Ph.D. in Chemical Education offered by the Department of Chemistry, but I didn't support the idea. Our existing programs were succeeding, so why rock the boat? The year after I left Purdue, that program was approved.

Purdue's Chemistry Department had a reputation for excellence in teaching long before I arrived. Derek Davenport, Hal Carter, Frank Martin, and Bob Livingston were recognized for their commitment to teaching and their interest in chemical education at the pre-college as well as the college level. The department has always had a cadre of outstanding teachers. Derek Davenport had an international reputation as a speaker and as a developer of creative experiments and demonstrations. To distinguish myself from Derek, whose reputation eclipsed my own, I pointed out that Derek is a chemist with an interest in education; I, on the other hand, am an educator with an interest in chemistry.

In 1970, I was promoted to Associate Professor with tenure, primarily on the basis of my teaching and work with high school teachers. But further promotion would almost certainly depend on recognition as a scholar. I was publishing science education research, presenting papers, and speaking at national meetings, but almost exclusively in education circles. While still teaching at the high school level, I had read some issues of the *Journal of Chemical Education*. Although I found some articles useful, there did not seem to be enough information for high school chemistry teachers about the theory of teaching and learning chemistry. At the time, the American Chemical Society had expensive dues, and there was no office or division for pre-college teachers. There was the Division of Chemical Education and the ACS *DivChemEd Examinations Institute*, but these seemed geared for college and university faculty. At the time, I thought I did not have anything to offer its members.

In the summer of 1974, Purdue University's Department of Chemistry was hosting the Great Lakes Regional ACS meeting and Derek Davenport, one of the organizers of the meeting, insisted that I submit a paper about my group's work in chemical education. By this time, I knew that proportional reasoning, "if…. then… therefore…" reasoning, and other components Jean Piaget (Flavel, 1963, 1977, 1985; Flavel, et al., 1993; Piaget, 1964, 1972) described as Formal Operations Reasoning were just as much a stumbling block for college freshmen as they had been for high school and junior high students. My graduate students and I had conducted some pilot studies and the results strongly suggested that college freshmen frequently failed chemistry because the reasoning underlying many chemistry concepts made no sense to them. I decided to report these findings and to explain Jean Piaget's theory, on which the findings were based, to chemists. Bob Benkeser, then Department Chair, moderated the session in which I presented, and he insisted that I submit my talk for publication in the *Journal of Chemical Education*.

I was still skeptical that chemists would have much interest in education theory; after all, it dealt with how students learn (or fail to learn) chemistry, not with chemistry content per se. Still, I knew that publication was important for tenure and promotion, so I submitted the piece to Tom Lippincott, then editor of the *Journal of Chemical Education*. I don't recall how many revisions Tom forced me to do before I got it right. I do recall how irritating it was that reviewers couldn't understand my clear, even erudite, prose! Derek had often complained about the educational jargon used by me and my friends who published in *Science Education* and *Journal of Research in Science Teaching*, two leading peer-reviewed journals for science education research at the time, but I countered that our language was no more "jargon" than "color," "charm," and similar words adopted by scientists and given technical meaning through research. Knowing that words that conveyed specific meaning to science educators were "jargon" to chemists made me no happier about having to change them! I'm now glad that I did.

The article, published in 1975 in the *Journal of Chemical Education*, was titled "Piaget for Chemists" (Herron, 1975), and it struck a responsive chord among college and high school chemistry teachers. The article cast light on a long-standing problem of widespread concern. For the first time, most everyone agreed that Piaget's theory offered an explanation as to why students had such trouble solving stoichiometry and gas law problems. If a student could not use proportional reasoning, then no matter how clear the lectures were, no matter how lucid the textbook was, the student was not able to solve problems that involved proportional reasoning. The underlying ideas of constructivism and stages of intellectual development that were presented in the article contributed to a national dialogue on how to improve chemistry instruction at both the high school and college levels.

Tom Lippincott wanted *Journal of Chemical Education* to serve high school teachers better. He published more articles about education theory and chemical education research, and I belatedly accepted his invitation to edit a column specifically for high school teachers. That column, *High School Forum*, was one of several features that Lippincott introduced to better serve high school teachers. Later, the *Journal* added the "Research: Science and Education" section featuring both chemistry experimental research and chemical education experimental research. To complement these changes in the *Journal*, ACS created an Education Office that focused on teaching and learning chemistry at all levels, and the ACS Examinations Institute enhanced its examinations program for secondary school courses, adding totally new kinds of examinations to assess conceptual understanding. At ACS national meetings, the Division of Chemical Education now has a full slate of symposia on teaching and learning chemistry. Every other summer, the Division of Chemical Education sponsors the *Biennial Conference on Chemical Education*, which attracts between 1000 and 1600 participants. There are regularly scheduled *Gordon Research Conferences on Chemical Education*. Articles describing and defining the field of chemical education research and scholarship have been published (Nurrenbern and Robinson, 1994; Bunce, Gabel, Herron, Jones, 1994; Bunce, Robinson, 1997; Herron and Nurrenbern 1999). Finally, in 2004, there are at least 23 chemistry departments in the USA that have hired chemical educators, and at least 16 have a Ph.D. program in chemical education (Bretz, 2004; Mason, 2001). These are just some indications that chemical education is coming of age. This book is another.

This book suggests how far chemical education has come in the past 30 years. Communication doesn't flow easily between faculty in psychology, education, and chemistry, but the authors of this book–virtually all faculty of chemistry–speak freely about cognition, teaching strategies, and cognitive style. Some chapters are jointly authored by chemists and colleagues in schools of education. The problems of teaching chemistry are being taken seriously, and we are accumulating knowledge about how to do it well.

Teaching and learning chemistry is no simple proposition. As Michael Kasha remarked during one of my doctoral examinations, "The mechanics of learning chemistry are surely as complex as a grand piano!" And they are. This book reflects that complexity. Those who are looking for a simple "how to" book will not find it here; those who seek a better understanding of the teaching and learning process will not be disappointed. For those who are serious about doing a good job in their classrooms, each chapter provides a wealth of references for you to continue your studies. It need not be an independent study course. Talk to colleagues within your department about teaching and learning; attend science education, physics education, or chemical education seminars on your campus; attend chemical education symposia at regional and national ACS meetings; read articles of interest to you in the *Journal of Chemical Education*; attend NSF-sponsored workshops on reforms in teaching and learning, participate in the *Biennial Conferences on Chemical Education*, and participate in a *Gordon Research Conference on Chemical Education*. The more you participate, the more you ask questions, the more you dialogue and challenge, the better your teaching will be.

Synopsis of the Chapters in this Book

Part I. Cognition. Common sense tells us there is no mechanism to transmit an idea, intact, from one brain to another, but we still act as though there is. Much as we know that "telling is not teaching," that knowledge has not altered the preeminence of lecture in college teaching. No chemist has the time or the inclination to become thoroughly versed in cognitive science, but any chemist who wishes to be an effective teacher needs to know some. In Chapter 2, Cracolice does an outstanding job of providing, in broad, clear strokes, enough information about how knowledge is constructed in brains to provide a framework for understanding how learning takes place. Read it carefully and reflect on the ideas as subsequent chapters show how the principles presented in Chapter 2 guide research and practice in chemical education.

Knowing how knowledge is constructed might be sufficient if everyone inherited the same DNA and lived through identical experiences. Neither being true, teachers face students who, happily, I think, demonstrate infinite differences. In recent years, researchers have described student differences in a manner that may give clues about how to maximize their learning. In Chapter 3, Bretz provides an overview of various schemes for describing learning styles, and she reviews research showing that learning styles do affect learning.

Because of differences in the ways students learn, many believe the best bet is to have students direct their own learning. Toward that end, science educators have sought manageable procedures that allow that to take place. Robert Karplus, the physicist who directed the *Elementary Science Study (ESS)*, proposed a three-step

Karplus

procedure for placing students at the center of learning. He called the procedure a learning cycle. Abraham has spent a long career using the learning cycle to guide inquiry in chemistry and researching its effectiveness. In Chapter 4, he makes compelling arguments for using inquiry as a teaching strategy. As Abraham says, there is no one *best* way to teach. But Abraham goes on to point out that "this cliché [should not] be incorrectly interpreted to mean that it makes no difference how one teaches."

Research clearly indicates that instruction based on sound teaching principles leads to better attitudes and better understanding on the part of students. Still, there are many strategies and tactics that are consistent with accepted learning theory. The learning cycle, or any strategy aimed at making science meaningful, will fail if students think their course is irrelevant, and they often do. However, understanding that it helps to make courses "relevant" is too simplistic; it is true, but there is more than meets the eye. As Donald Wink (Chapter 5) points out, if we are to make chemistry relevant to students, we must attend to the particular student(s) we are trying to teach, we must pay attention to our role as mediator of instruction, we must use the right kind of instructional materials, and we must ensure that assessment of learning is consistent with instruction. Wink uses relevant chemistry materials to illustrate some of these problems and then provides concrete suggestions for doing better.

Attempting to make courses relevant is but one tactic used to make chemistry comprehensible. In Chapter 6, Bodner, Gardner, and Briggs discuss models and modeling extensively. They define models, discuss various kinds of models, and show how models are effectively used in teaching. Their focus, however, is a rationale for emphasizing modeling itself–how models evolve, their limitations, and why we need them.

Chemists use many kinds of models, and the atomic model plays a central role in our thinking, as do the symbols used to represent atoms and molecules. Chemists think in three worlds: the macroscopic world, where solids sometimes turn into liquids; the molecular world, where water molecules attach to and insinuate themselves between solute particles; and the symbolic world, where representations like $H_2O(l)$, $C_{12}H_{22}O_{11}(s)$, and $C_{12}H_{22}O_{11}(aq)$ convey, to those who are savvy, information gleaned from other realms. Chemists are so accustomed to switching their thoughts from one of these worlds to another that they cannot imagine that any intelligent person doesn't do the same. Alas, it isn't so, as Dorothy Gabel points out in Chapter 7. If you are not familiar with the large body of research documenting student difficulties in chemistry and the contribution that our facile movement from the macroscopic to the microscopic to the symbolic makes to those difficulties, Gabel's chapter will prove informative.

Part II. Teaching Strategies. The first part of this book focuses on how students think about chemistry and offers an overview of some of the main educational theories that have proven effective. The second part examines strategies that you can use to help students think more like chemists and help them understand the concepts and principles of chemistry. Many chemists are surprised to learn that analogies can confuse students. The common understanding is that analogies clarify abstract ideas, and indeed, they might. However, what appears on the surface to be straightforward turns out to be exceedingly complex, as Orgill and Bodner tell us in Chapter 8.

Given the multitude of variables that affect teaching and learning, it should be no surprise that the effectiveness of an analogy depends on (a) student familiarity with the analog, (b) analytical skill of the students, (c) how carefully teachers focus on connections between analog and target concepts, (d) difficulty of the target concept, and (e) other variables. Teachers looking for a pat formula for effective use of analogies will not find it. Classroom environments are too complex for teaching to be algorithmic. But research on analogies provides insight that can guide resourceful teachers. If you ever use analogies (and who doesn't?), you will want to read this chapter and think about its implications.

Teaching students to solve mathematically-based word problems that appear as end-of-chapter exercises in textbooks is a formidable challenge. It is this challenge that Bunce addresses in Chapter 9. There is a plethora of research on problem solving, and much of it addresses the kind of exercise described by Bunce. If you aren't familiar with how students typically approach these exercises, Bunce will bring you up to speed. Bunce illustrates typical student behavior, shows that students can learn to apply the skills used by experts, and explains why the most common teaching tactics don't do the job.

Even the strategies that successfully teach students to solve textbook problems may have little impact on their ability to complete novel tasks, such as those encountered in any research environment. The vast body of research related to this kind of problem solving will be addressed in the next book in this series.

All of the teaching techniques described here have one thing in common: they represent strategies to engage students in thinking about chemistry ideas. Far too much of schooling involves students in mindless exercises that produce written products to be graded but lead to little or no change in conceptual understanding. That is certainly not the instructor's intent, but it is what usually happens.

The negotiation of meaning in groups is an effective way to consolidate knowledge, and a variety of group learning activities have been used in chemistry. Unfortunately, placing students in groups does not guarantee that negotiation of meaning takes place. Research in many disciplines has produced reliable evidence about conditions needed for group learning. Cooper (Chapter 10) does a nice job of summarizing that research, describing applications of group work in chemistry, and citing references where the reader can get detailed information about how to use cooperative group learning.

You will notice that the authors of this book seldom recommend using the teaching strategy or tactic that they describe in isolation. Group work, for example, is employed with concept mapping and the writing heuristic described in Chapters 11 and 12. Concept maps are pictorial representations of interrelationships among concepts, and they can be used in many ways. Nakhleh and Saglam (Chapter 11) describe how concept maps can be used in the laboratory to achieve the same goals described by Hand and Greenbowe in Chapter 12, but that is just one application they describe. Concept maps can be used as group work in lectures and recitations and for assessment as well. If you are unfamiliar with concept maps, you will want to look at this chapter to see potential uses in your courses.

Can students learn chemistry by writing explanations for chemistry events? Hand and Greenbowe (Chapter 12) believe the answer is "yes" … *if* you and your students implement the appropriate techniques of guided-inquiry, group work, and the Science Writing Heuristic. An excellent place to start is in the chemistry teaching laboratory. Like the learning cycle described by Abraham in Chapter 4, the Science Writing Heuristic provides structure for laboratory work without allowing students to complete the work mechanically. Students must think about what they will do in the laboratory and why. Once they have data in hand, they must decide what it means, and in pooling their data and interpretations with that generated by classmates, they look for patterns and rationalize inconsistencies that crop up. In short, they construct meaning, the essence of all learning.

Research chemists will identify with the Team Learning (Chapter 13) described by Varma-Nelson and Coppola. There is a strong similarity between teams and research groups. In addition to Johnson and Johnson's work on cooperative learning cited by Cooper in Chapter 10, Varma-Nelson and Coppola draw on Vygotsky's social constructivism (described by Cracolice in Chapter 2), Palincsar and Brown's work on reciprocal teaching, and studio instruction to guide their work. Strong empirical evidence for the effectiveness of team learning, coupled with the similarity to research groups, make this an attractive teaching model for chemists who want to try a new approach in teaching but are unfamiliar with (or perhaps skeptical of) theory and practices that have emerged from research in cognitive science and education.

Part III. Learning With Technology Chemists who are intimidated by learning theory may be more comfortable with some of the ideas presented in this section. Chemists are familiar with technology and often find it easy to apply in their classroom. In Chapter 14, Pienta and Amend explain how to get started with electronic data collection (EDC) in the laboratory, and they provide an excellent review of previous research on the effectiveness of laboratory instruction. EDC can be employed just to save time, but Pienta and Amend favor EDC because it enables them to emphasize inquiry. In Chapter 4, Abraham explains why. Abraham has spent a long career using inquiry in teaching chemistry and researching its effectiveness. He makes compelling arguments for using inquiry as a teaching strategy.

Another point of entry for chemists who are shy about delving into newfangled pedagogy is found in Reeves and Ward's chapter on the use of wireless technology (Chapter 15). Most chemists are technologically literate, and many of Reeves and Ward's suggestions are for using technology to improve lectures and other familiar modes of instruction.

In Chapter 16, Roy Tasker describes another use of technology that can be applied in conventional instruction. Atomic theory is a powerful organizer of chemical information, but only if one is able to visualize atoms and molecules. Unfortunately, learning to think about particles that cannot be perceived by any of our senses is no simple task, and poorly conceived animations often instill misconceptions rather than clarify accepted models. You will see that Tasker has paid careful attention to research on learning, has applied principles derived from that research to his work, and has gone through repeated iterations to eliminate potential misconceptions and maximize the usefulness of his materials. Tasker has formed partnerships with commercial entities that make his materials widely available.

Conclusion

I have often stated this professional goal: "Before I die, I would like to see somewhere on Earth an institutional structure where individuals on the cutting edge of research on teaching and learning and individuals on the cutting edge of research in chemistry are working together and talking to one another." I don't think that goal has been reached, but I'm encouraged that it will be. This book does not include all topics serving as the foundation of science education theory and practice, but it does include most topics relevant to chemists teaching any level of chemistry.

Increasingly, chemists in traditional fields are recognizing that chemistry education is a legitimate branch of the discipline and that research in that branch provides useful information. We are, it seems, where medicine was around 1910, often cited as when a person entering a hospital first had a greater than 50-50 chance of coming out alive!

I have said that chemistry education is coming of age; I have not said it has arrived. Some chemical educators will continue to complain like Roger Dangerfield, "I don't get no respect," because there are incompetent people who claim to be chemical educators, just as there are charlatans who claim to be doctors. Other chemists are beginning to realize that is an aberration rather than the rule. In the departments where the authors of this book reside, interesting conversations take place about teaching and learning. Colleagues are interested in what they know. Hopefully, the authors will someday have the same giddy feeling I had while reading their chapters. Some readers of this book will be old enough to understand: I felt like the kid in the old television commercial proudly exclaiming, "We used Shake 'n Bake and I helped!"

Recommended Readings

Bunce, D., Gabel, D., Herron, J. D., Jones, L. (1994). Report of the Task Force on Chemical Education Research of the American Chemical Society Division of Chemical Education. *Journal of Chemical Education*. 71(10), 850.

This article discusses the scholarship of teaching and the scholarship of discovery and argues for the legitimacy of chemical education faculty in chemistry departments.

Bunce, Diane M.; Robinson, William R. (1997). Research in Chemical Education–The Third Branch of Our Profession. *Journal of Chemical Education*. 74(9), 1076–1079.

This article compares chemical education research to chemistry research.

Herron, J. D.; Nurrenbern, S. C. (1999). Chemical Education Research: Improving Chemistry Learning. *Journal of Chemical Education*. 76(10), 1353–1361.

This article discusses how specific chemical education research articles have helped improve students' understanding of difficult chemistry concepts or have improved chemistry teaching.

References

ASTM A90/A90M01, "Standard Test Method for Weight (Mass) of Coating on Iron and Steel Articles with Zinc or Zinc–Alloy Coatings," in (2004) *Annual Book of ASTM Standards*. West Conshohocken, PA: ASTM International.

Bretz, S. L. Division of Chemical Education (ACS) Committee on Chemical Education Research: Masters (M.S.) and Doctoral (Ph.D.) Programs in Chemical Education. Available online at http://www.as.ysu.edu/~slbretz/cer/programs.htm Last accessed August 2004.

Bunce, D.; Gabel, D.; Herron, J. D.; Jones, L. (1994). Report of the Task Force on Chemical Education Research of the American Chemical Society Division of Chemical Education. *Journal of Chemical Education.*71(10), 850.

Bunce, D. M., and Robinson, W. R. (1997). Research in Chemical Education–The Third Branch of Our Profession. *Journal of Chemical Education*. 74(9), 1076–1079.

Flavel, J. H. (1963). The developmental psychology of Jean Piaget. Princeton, NJ: Van Nostrand.

Flavel, J. H. (1977). Cognitive development. Englewood Cliffs, NJ: Prentice-Hall.

Flavel, J. H. (1985). Cognitive development (2nd ed.) Englewood Cliffs, NJ: Prentice-Hall.

Flavel, J., Miller, P., and Miller, S. (1993). Cognitive development (3rd ed.) Englewood Cliffs, NJ: Prentice Hall.

Herron, J.D. (1975). Piaget for Chemists: Explaining what "good" students cannot understand. *Journal of Chemical Education*. 52(3), 146–150.

Herron, J. D., and Nurrenbern, S. C. (1999). Chemical Education Research: Improving Chemistry Learning. *Journal of Chemical Education*. 76(10), 1353–1361.

Mason, D. (2001). A survey of doctoral programs in chemical education in the United States, *Journal of Chemical Education*. 78(2), 158–159.

Nurrenbern, S. C., andRobinson, W. R. (1994). Quantitative Research in Chemical Education. *Journal of Chemical Education.,* 71(3), 181–183.

Piaget, J. (1964). Development and learning. *Journal of Research in Science Teaching*. 2, 176–186.

Piaget, J. (1972). Intellectual evolution from adolescence to adulthood. *Human Development*. 15, 1–12.

2

How Students Learn:
Knowledge Construction in College Chemistry Courses

Mark S. Cracolice
Department of Chemistry
The University of Montana

Abstract

Theories and philosophies of the mechanism of human learning and results from experiments based on those foundational pillars should serve as the basis of all instructional design, but repeating the past and teaching as one was taught is the most common method of instruction in most college chemistry courses. If chemistry instructors are to significantly improve the quality of course materials and pedagogical approaches, they must first understand the fundamentals of how students learn. The key idea presented in this chapter is that the highest-quality learning and intellectual growth occurs when students make observations of natural phenomena and are subsequently challenged to construct conceptual understanding from the raw information. Concepts learned in specific chemistry contexts can then be generalized to broader contexts that can apply across disciplines. The theoretical foundations of this conclusion come from work by Piaget, Vygotsky, and Ausubel.

Biography

Mark Cracolice is professor and chair of the Department of Chemistry at The University of Montana. His prior administrative position was as the Director of the Center for Teaching Excellence at The University of Montana. Cracolice teaches introductory chemistry, general chemistry, and graduate courses in chemical education. Additionally, he supervises a Ph.D. program in chemistry, cognitive psychology, and education with an emphasis in chemical education research. The primary focus of his research group is to develop models of chemistry curricula that are consistent with models of human learning. He also works with local high school teachers to design chemistry curricula that promote the development of higher-order thinking skills. Cracolice is co-author of a number of journal articles in chemical education, an introductory chemistry textbook, and publications on Peer-Led Team Learning.

Introduction

Knowledge of what is known about human learning is central to mastering the craft of teaching—or is it? Few college chemistry instructors have experienced meaningful formal or individualized instruction in educational psychology or cognitive science, and such an education is not required for an advanced degree in a science discipline. The generally accepted view is that expertise in chemistry research is the primary training needed for a college chemistry instructor. An advanced researcher is assumed to have acquired the tools of the teaching trade simply by being an experienced student. This tradition leads to a cyclical process of teaching as one is taught; it is a cycle where no improvement is possible. The fact that you are reading this book indicates that you recognize the potential pitfalls of this cycle.

Unfortunately, this historical standard—or lack of it, if you will—leads to two prevalent misconceptions held by chemistry instructors:

1. The primary goal for student achievement in a course is increased content knowledge in the discipline.
2. Students should be able to learn effectively from traditional high-quality content delivery tools such as textbooks, lectures, laboratory manuals, and multimedia presentations.

Conclusion: Organize the curriculum according to principles of the discipline (e.g., smallest to largest particles, elementary to complex systems, etc.) and deliver this information with beautiful four-color textbooks and PowerPoint presentations, and *voilà*, students will learn! Alas, this conclusion couldn't be more wrong.

In this chapter, we will explore why these misconceptions are commonly held, and we will examine the theoretical and experimental bases that can be applied to improve the quality of instruction in science classrooms. Subsequent chapters will describe applications of these theories. Before we begin an exploration of the current state of knowledge about human learning, however, it will be beneficial to look to the past to understand how we arrived at this point in history. Specifically, we will take a brief look at the progress in curriculum materials for general chemistry. The evolution of the organization of general chemistry curricula reflects how a wide spectrum of science courses has been taught in the past century.

In most of the twentieth century, general chemistry was taught on the basis of content-centered designs. The content of chemistry itself was the basis of the organizational structure of the textbook. At first, the principles of descriptive chemistry served as a logical systematizing structure. *Smith's College Chemistry* (Kendall, 1935) was perhaps the most popular general chemistry textbook in the early part of the century. Its chapters included Oxygen (Chapter 6), Hydrogen (Chapter 8), Hydrogen Chlorides and Sodium Hydroxide (Chapter 13), and Chlorine (Chapter 14).

A revolutionizing change in general chemistry texts came with the publication of books with concepts-centered designs. Textbook titles such as *Principles of Chemistry* (Sanderson, 1963), *Chemical Principles* (Masterton and Slowinski, 1966), and *Chemistry: Principles and Properties* (Sienko and Plane, 1966) reflected this change. In the 1950s and 1960s, the organizing theme changed from chemical reactions to chemical principles. Nonetheless, the focus of the general chemistry curriculum remained on content. It is also important to note that even a principles-centered design is based on an expert's mental organization of the discipline, not a student's. (We'll discuss this in more detail later.) A guiding philosophy with this organization is that if the method of delivery is perfected, students will understand. However, as we enter the 21st century, we find that as we add color photography, CD-ROMs with animations and video, web supplements, etc., we have not necessarily improved student *learning*—even with the concepts-centered curriculum strategy.

Studies have found no significant learning gains from delivering the same content via web-based homework or using PowerPoint-based lectures when compared with equivalent control groups (Bonham, Deardorff, and Beichner, 2003; Cole and Todd, 2003; Szabo and Hastings, 2000). However, it is important to point out that this is not intended to be a sweeping statement accusing all technology of being ineffective in improving student learning. The point is that high-quality delivery of content does not necessarily lead to learning gains. Richard Clark (1983) argues that instructional media "… are mere vehicles that deliver instruction, but do not influence student achievement any more than the truck that delivers our groceries causes changes in our nutrition" (p. 445). However, Robert Kozma (1991, 1994) and Jack Koumi (1994) counterargue that media can have a unique effect on learning, but many research studies are based on poor-quality media, and therefore, additional research should be conducted. Clark (1994) continues to be unswayed by the counterarguments, claiming that media will never be the *cause* of learning and will actually be harmful by increasing the cost of instruction.

Another educational revolution occurred simultaneously in the 1950s and 1960s. This revolution was the result of an emerging focus on the psychology of learning. A radical new idea began to take root: instead of designing curricula around the principles of a discipline, shape the curriculum to be consistent with how people learn. At this time, behaviorism was the dominant paradigm in educational psychology. Behaviorists studied only the observable, quantifiable aspects of an organism's behavior, treating the underlying cognitive processes as an unknowable black box (O'Donohue and Kitchener, 1998; Skinner, 1974). Thus, new curriculum designs emerged that attempted to break complex learning tasks into a series of simple steps, often accompanied with positive reinforcement for the student in the form of frequent feedback. A number of chemistry instructors embraced the psychology-centered revolution in the form of Personalized System of Instruction (PSI) courses (Keller, 1968). PSI courses feature an emphasis on mastery of the material and self-pacing. Students are allowed to have as much time as they need to demonstrate an understanding of course content, but they have to show mastery in a topic—usually in the form of scoring 90% or more on exams—before they are allowed to move to the next topic. Exams are repeated until the mastery standard is achieved. This form of instruction gained popularity from about 1970 to 1985, but has been largely dormant since (Cracolice and Roth, 1996). PSI courses fell out of favor largely because of the daunting initial workload required to produce multiple forms of equivalent exams for each textbook chapter throughout the term. Additionally, the self-paced feature of the course design leads to significant student procrastination and administrative difficulties (such as carrying incomplete grades from term to term).

Behavioral-centered designs gave way to cognitive-centered designs (Parkin, 2000; Sternberg, 2002) in the late 20th century. Cognitive psychology is based on hypothesizing about the mental process that causes behavior; behaviorism focuses only on the behavior itself. Characteristics such as motivation, desire, and other mental states that drive behavior are excluded in behaviorism, but they are included in cognitive models. Although curriculum designs derived from behaviorist principles were good for learning problem-solving algorithms and memorization of facts, instructors discovered that students could get the right answer to many problems with little to no conceptual understanding (Bunce, Gabel, and Samuel, 1991; Gabel, Sherwood, and Enochs, 1984; Herron and Greenbowe, 1986; Nurrenbern and Pickering, 1987). When students were stretched slightly beyond the algorithms in the curriculum design, they failed dismally. We will therefore discuss cognitive-centered curricula in detail throughout the remainder of the chapter.

With the advent of instructional designs based on learning principles, a new way of looking at the ultimate purpose of formal education began to evolve. Of course, "new" means different from what immediately preceded this idea, but this "new" purpose has a history that dates back to the origins of recorded civilization. It is the root of the liberal arts tradition in college education. "The purpose which runs through and strengthens all other educational purposes—the common thread of education—is the development of the ability to think" (Educational Policies Commission, 1961). This central purpose is one that college chemistry instructors too often neglect. Our task is not only to teach the content of our disciplines, but also to facilitate the improvement of our students' intellect. Moreover, science courses are an ideal environment in which to promote intellectual development. In traditional course designs, most students do not learn to think. They process course content using their present thinking abilities. It is our responsibility as science instructors to change our curricula to help our students develop their intellectual skills, so that they can understand college-level chemistry, as we simultaneously teach the content of our disciplines (Deming and Cracolice, 2004). Throughout this book, methods to promote intellectual development will be provided.

Constructivism

While the educational community debates the precise definition of the theory or philosophy of learning known as constructivism (Scerri, 2003), our purpose here is simply to put forth a straightforward description of how people learn. The process by which a person acquires knowledge begins with input from the environment, as detected by the senses. Each individual constructs his or her own knowledge from data obtained by the senses and its interaction with existing knowledge; this is the essence of the constructivist theory of learning (Bodner, 1986; Bringuier, 1980). This implies that knowledge is the ultimate personal possession. The sum of my knowledge is unique and necessarily different from yours. You cannot transfer knowledge, fully intact, directly from your brain to mine.

To further understand constructivism, it is important to contrast it with alternative philosophies about human learning. One that most science instructors will readily reject is the nativist doctrine, which states that the mind produces knowledge that is not derived from external sources (Chomsky, 1975; Chomsky, 1980; Piatelli-Palmerini, 1980). Nativists believe that knowledge—such as language—is possessed at birth, unfolding as we mature. Chomsky argued that the development of language could not be explained by behaviorism and that an innate mechanism accounted for features common to all languages. For example, the sentence *Visiting relatives can be a nuisance* has more than one meaning. Are the relatives visiting you or are you visiting them? Chomsky argued that the deep structure representing the meaning of such a sentence, which goes beyond simple grammatical structure, provided evidence that there must be an innate capability for language acquisition and its associated abstract structure. Certainly, there are innate mental capabilities in place in human brains, but novel construction of knowledge occurs, so there must be something more than these inborn capabilities (Lawson and Staver, 1989; Quartz, 1993).

An alternative supposition is that knowledge can be directly transferred from external sources. In fact, this is the theory of learning being applied when an instructor lectures because the implicit assumption is that knowledge can be transferred directly from the mind of the lecturer to the minds of the students. This empiricist doctrine sees knowledge as something that is obtained through careful observation of the natural world (Matthews, 1992; Matthews, 1993; Nola, 1997). Every experienced instructor knows, however, that even our best students don't always understand chemistry concepts, even when we teach them very meticulously.

To be as practical as possible, and to avoid a philosophical debate that focuses on academic nuances and subtleties, it is most likely that elements of constructivism, nativism, and empiricism are all correct. The complexities of human cognition cannot be classified into a simple category, just as chemical bonds cannot be simply classified as entirely ionic, covalent, or metallic. But, as with the case of sodium chloride having mostly ionic bonding, a constructivist theory of knowledge acquisition is the most applicable in the chemistry classroom: people construct meaningful scientific knowledge for themselves. The use of constructivism as a model for learning in the science classroom provides us with a pragmatic theory base from which we can design effective curricula that allow students to learn scientific concepts. However, translating the science of learning into the art of teaching will always present a challenge. Nonetheless, the overarching principle to be utilized is simple: meaningful scientific knowledge is acquired by an individual's active mental analysis of information, not by direct transmission from a person or textbook (Bodner, 1986; Bodner, Klobuchar, and Geelan, 2001; Herron, 1996; Piaget 1977a, 1977b; Resnick, 1980; von Glassersfeld, 1981, 1995; Wittrock, 1974). A brief look at the key ideas of some of the primary architects of the constructivist theory will enhance this definition.

Piaget

Jean Piaget (1896–1980) was a Swiss psychologist whose work has had more influence on science education than the work of any other individual. While Piaget's studies were very extensive, he largely studied children learning in the absence of formal instruction; so while he helps us understand students' thinking, he largely neglects a direct statement about the role of the instructor. We therefore utilize Vygotsky's theories in addition to Piaget's theories to gain a more complete picture of constructivism. Piaget's theories will be discussed in greater detail later in this chapter.

Vygotsky

Lev Vygotsky (1896–1934) was a Russian psychologist whose brief life and residence in the former U.S.S.R. initially limited his contributions to our understanding of human learning, but his influence continues to grow. Vygotsky's key contribution to constructivism was in emphasizing the importance of the role of the instructor in the learning process. His most well-known theoretical construct is called the Zone of Proximal Development (ZPD): "The discrepancy between a child's actual mental age and the level he reaches in solving problems with assistance indicates the zone of his proximal development" (Vygotsky, 1986). Vygotsky advocated the importance of teaching within a student's ZPD, providing problems that are just beyond a student's current problem-solving ability. The role of the instructor is to give hints and guidance about how to solve the problem. In this way, a student is drawn towards a higher level of problem-solving ability. Note, in particular, the role of the instructor. A person, or materials created by a person, must be able to assess an individual's ZPD, provide problems just beyond the student's present capabilities, and then give assistance in solving those problems.

In the chemistry classroom, an instructor can teach within a student's zone of proximal development when tutoring students by probing for a student's understanding of underlying elementary skills and asking questions about the connections among these simpler skills. For example, if a student does not understand how to solve a mass-to-mass stoichiometry problem, a Vygotskian approach would be to ask the student about their understanding of molar mass, equation writing and balancing, and the other fundamental steps that make up the problem. If any of these were not understood, the student would be given assistance on learning that underlying procedure. If all fundamental steps were understood, the problem then must lie in connecting the steps. If connectivity is the learning barrier, then Vygotsky would suggest providing hints to the student and helping him or her build knowledge of how to make connections via a Socratic dialogue, rather than just showing the connections.

Vygotsky also discussed another key idea with profound implications for curriculum design in science courses. He believed that the basic psychological variable used to mentally organize a system of concepts is the degree of generality. Since concepts themselves are generalizations, the psychological nature of the relationships among concepts is a relationship among generalities. It is the absence of such an organizational system that leads to the development of alternative conceptions and other such troubles with knowledge construction. On the other hand, the critical event in reorganizing knowledge to gain a higher level of conceptual understanding is the shifting from a lower level to a higher level of generalization. We will further discuss this point later in the chapter, but in the interim, a brief example will be illustrative.

A typical general chemistry curriculum design introduces mass stoichiometry in the stoichiometry chapter, solution stoichiometry in the aqueous reactions chapter, thermochemical stoichiometry in the thermochemistry chapter, gas stoichiometry in the gases chapter, and electrochemical stoichiometry in the electrochemistry chapter. Initially, a student mentally organizes each of these stoichiometry concepts as they were taught: associated with the overarching principle of the textbook chapter. Some students will recognize the utility of reorganizing their knowledge on the basis that all of the stoichiometry concepts have the pattern: measurable quantity (given) to moles (given) to moles (wanted) to measurable quantity (wanted). The reorganization of knowledge raises the student's understanding to a higher level. An even higher level of understanding occurs when the generalized stoichiometry concept is further generalized to the concept of proportional reasoning. Solving a stoichiometry problem is, in essence, converting among a number of direct proportionalities. At this point, the student has abstracted the stoichiometry concept to its highest level of generalization, a formal thinking skill.

Ausubel

David P. Ausubel (1918–) is an American psychologist whose two major works were primarily brought to the attention of the educational community by Joseph Novak (Ausubel, 1963; Ausubel, Novak, and Hanesian, 1978). Ausubel himself wrote an oft-quoted passage that provides a summary of his major contribution to constructivism: "If I had to reduce all of educational psychology to just one principle, I would say this: The most important single factor influencing learning is what the learner already knows. Ascertain this and teach him accordingly" (Ausubel, Novak, and Hanesian, 1978, p. iv). Lawson (1995) reported on a personal communication with Ausubel in which the term "knows" in the Ausubel quote was clarified to include *both* declarative and procedural knowledge. Declarative knowledge is "knowing that"—in other words, knowing facts, such as the fact that BMW stands for Bavarian Motor Works. Procedural knowledge is "knowing how"—in other words, knowing how to do something, like knowing how to drive a BMW. Ausubel's main contribution emphasizes that learning occurs through the modification of preexisting knowledge. New knowledge—knowledge of something or of how to do something—is profoundly influenced by what the learner already knows.

There are a few important ways to apply Ausubel's theory to the teaching of chemistry. One is to use the advanced organizer, a curriculum element that can be used to bridge what a student already knows with what they are about to learn. Ausubel recommends that the most general concept be presented first, and then the more specific. Additionally, previously taught concepts should be continually cross-referenced during the introduction of new concepts. Another application is Gowin's Vee, a V-shaped diagram that can be used to formalize the relationship among what is already known and what is to be learned (Novak and Gowin, 1984). The vee diagram is constructed so that one side of the vee lists the conceptual aspects of an activity, while the other side lists the methodological aspects. At the point of the V is the event or object to be studied to answer the research question. Finally, concept mapping is derived from Ausubel's theory. This is covered in detail in Chapter 11 and in part Chapter 3, so we do not discuss it here.

Piaget's Theories

Piaget's theories of intelligence and quality of thought are at the core of the constructivist theory and therefore deserve an extended discussion. For those who are interested in an outstanding article that gives a more thorough introduction to Piaget, see Herron (1975).

Knowledge

Perhaps the most important idea regarding scientific knowledge construction that comes from Piaget's theory of intelligence is that of a *scheme*. Piaget defined a scheme as "the structure or organization of actions which is transferred or generalized when this action is repeated in similar or analogous circumstances (Piaget and Inhelder, 1969, p. 11). In other words, a scheme is an arrangement of actions that can be applied in multiple circumstances. Piaget also used the term *operatory scheme,* or "general abstract schemes, i.e., concepts taking the form of classes or relations" (Piaget, 1962, p. 243). Therefore, a scheme is a way of organizing knowledge, either physical in the sense of how to do something bodily, or mental, in the form of knowing a concept.

Piaget postulated that schemes are changed through a dynamic process of equilibration, where new information is compared with existing schemes; he also postulated that when a mismatch occurs, the scheme may be modified. "Every assimilatory scheme has to be accommodated to the elements it assimilates, but the changes made to adapt it to an object's peculiarities must be effected with loss of continuity. This postulate indicates that modifying a scheme must destroy neither its closure as a cycle of interdependent processes nor its previous powers of assimilation" (Piaget, 1985, p. 6). It is the interaction between assimilation, the incorporation of new information into scheme, and accommodation, the change in a scheme that occurs because of the new information, which leads to revision of schemes. Equilibration is essentially a process of self-regulation, which is explored further later in the chapter.

An example of how a student constructs knowledge, according to Piagetian theory, can be illustrated by considering the following exercise: *What caused the water to rise?* Students are instructed to place a burning candle in the center of a pan of water, and then invert a glass cylinder over the candle. The flame goes out and the water rises in the cylinder. Students are asked to explain the cause of the water rise. The typical answer is that oxygen is consumed, creating a vacuum, which sucks the water into the cylinder. As students observe these events, they are *assimilating* the data into their pre-existing mental structures which typically allow for matter to be destroyed and for "sucking."

Students are then instructed to repeat the experiment with multiple candles. They observe that the greater the number of candles, the higher the water rise. This contradicts the "oxygen is consumed" explanation posed earlier. Students go into *disequilibrium.* When they learn that more candles cause more air to escape from the cylinder, they learn that air pressure is the true cause of the phenomenon, and they revise their mental scheme, *adapting* to the new knowledge. With additional experience, this revised scheme is *organized* with respect to other schemes related to the air pressure phenomenon. Learning has occurred.

Stages of Quality of Thought

Another essential idea that originates with Piaget is that the quality of thought in individuals' progresses is a series of stages which likely correlate with stages in brain development. As a child's brain develops, the ability to think undergoes a step-wise progression, ideally characterized by a few years at any particular stage, followed by a relatively rapid transformation into the subsequent stage. Piaget said, "We must note that each level is characterized by a new coordination of the elements provided—already existing in the form of wholes, though of a lower order—by the processes of the previous level. Each of the transitions from one of these levels to the next is, therefore, characterized both by a new coordination and by a differentiation of the systems constituting the unit of the preceding level" (Piaget, 1950, pp. 151–152).

Piaget proposed four stages of development: sensorimotor, typically from birth to 1.5 years of age; preoperational, until the age of about 7 years; concrete, beginning at approximately age 7; and formal, which can develop as early as age 11. High school and college instructors need not be concerned with the sensorimotor and preoperational stages, so they will not be addressed in this chapter. Our focus will be on the contrast between the concrete and formal stages.

The quality of thought at the concrete stage is characterized by the need to see physical objects or models of objects. Only then can objects be imagined and thought about. Another characteristic of concrete thinking is that it is dominated by context dependent cues and intuitions (Lawson, Abraham, and Renner, 1989). It is largely intuitive. The ability to think at the concrete level is typically measured by assessing the ability to perform the tasks in Table 1. For example, conservation of displaced volume can be assessed by having the subject observe the water displacement in a graduated cylinder when an aluminum weight is lowered into the water. The subject is then asked to state the volume of water that will be displaced by a lead weight of the same volume. One who has concrete thinking ability will correctly understand that the volume of the object is the critical variable, not its weight.

Concrete Thinking Skill	Description
Conservation of number	The number of objects in a group remains the same no matter their arrangement.
Conservation of liquid amount	The amount of liquid remains the same no matter the size or shape of the container in which it is held.
Conservation of solid amount	The amount of solid remains the same no matter its shape.
Conservation of area	The area uncovered remains the same when equal numbers of objects of equal area are placed on a two-dimensional surface, no matter whether the objects are placed together or apart.
Conservation of length	The length of objects is the same no matter their shape or position.
Conservation of weight	The weight of a solid remains the same no matter its shape.
Conservation of displaced volume	Displaced volume is equal to the volume of the displacing object, no matter its density.
Ordering	An observable property can be used to establish an order among objects, events, etc.
Correspondence	Different objects can be related in number, size, amount, etc.
(Simple and multiple) classification/class inclusion	Objects can be classified into one- or two-dimensional categories; a class can exist within another more inclusive class.

Table 1. Summary of concrete thinking skills (Bybee and Sund, 1990; Marek and Cavallo, 1997; Piaget and Inhelder, 1969).

Formal thought is characterized by the ability to imagine unobservable entities. It is not context bound. A formal thinker can read a written description of an ion in solution and form a mental model of the solution. Piaget wrote, "As of eleven to twelve years, formal thinking becomes possible, i.e., the logical operations begin to be transposed from the plane of concrete manipulation to the ideational plane, where they are expressed in some kind of language (words, mathematical symbols, etc.), without the support of perception, experience, or even faith…. Formal thought is 'hypothetico-deductive,' in the sense that it permits one to draw conclusions from pure hypotheses and not merely from actual observations. These conclusions even have a validity independent of their factual truth" (Piaget, 1967, pp. 62–63).

Formal thinking can also be thought of as reflective thinking, where one consciously reflects on their thinking (Lawson, Abraham, and Renner, 1989). Table 2 gives a number of formal thinking skills that are often used in evaluations of formal thinking ability. Consider how these skills are essential for understanding the content of a science course. Control of variables is at the core of the scientific method. Ratio and proportion are used throughout the physical sciences. Formal models are at the heart of chemistry. Students must have the ability to think formally to succeed in almost any high school or college science course.

Formal Thinking Skill	Description
Control and exclusion of variables	Holding n independent variables and one dependent variable in mind, and considering the possible effects of each independent variable on the dependent variable.
Classification	The processes of understanding the possible ways that a classification may be carried out,

	understanding that any classification operation is part of a hierarchical system, selecting different criteria for different purposes, and understanding that one particular criterion does not necessarily allow prediction of others.
Ratio and proportion	Ratio: $y = mx$ (as x goes up, so must y); Proportionality: Comparison of two ratios.
Compensation and equilibrium	Compensation: $yx = m$ (as y goes up, x must come down); equilibrium: $ab = cd$.
Correlation	Determination of correlation among variables.
Probability	Simple sampling procedures; acceptance of the probabilistic nature of natural relationships
Formal models	Model: representation of something else; working model: has different parts which move and which hold the same relationships to one another as in the real thing; formal model: a working model in which the moving parts are abstract entities which have to be imagined.
Logical reasoning	The ability to analyze the combinatorial relations present in information given.
Hypothetico-deductive reasoning	The ability to formulate and test alternative hypotheses against given data.

Table 2. Summary of formal thinking skills (Adey and Shayer, 1994; Inhelder and Piaget, 1958).

McKinnon and Renner (1971) described the reasoning abilities of students entering their freshman year at a large public university: 50.4% of students were classified as concrete, 24.4% as transitional, and 25.2% as formal. In this study, 75% of students entering college were not fully formal reasoners. Subsequent studies have shown approximately the same results (Lawson et al., 2000a; Valanides, 1999).

The dislocation between the need for formal thought in science courses and the formal thinking abilities of students presents nothing short of an educational crisis. Its importance cannot be overstated. Given that concrete thinkers cannot learn formal concepts (Lawson and Renner, 1975), a course with 75% of students who are not fully formal yields at least 75% who do not understand the course content after completing the course. Since concrete thinkers typically memorize their way through formal concepts, almost no course material will be part of a student's permanent knowledge. It is *imperative* that science instructors design curricula to facilitate the development of students' thinking abilities.

These formal thinking abilities can be measured with paper-and-pencil instruments such as the Classroom Test of Scientific Reasoning (Lawson 1978, 1987), the Test of Logical Thinking (Tobin and Capie, 1980a, 1980b), or the Group Assessment of Logical Thinking (Bunce and Hutchinson, 1993; Roadrangka, Yeany, and Padilla, 1982), all of which are based on traditional Piagetian clinical interview tasks. For example, Piaget utilized a pendulum task, where the length of the string, mass of object tied to the string, height from which the mass was initially dropped, and the force of the initial push could potentially affect the frequency of the oscillation. The subject was allowed to manipulate the apparatus, and in order to be classified as fully formal on the separation and exclusion of variables thinking skill, they would have to systematically work with the apparatus to come to the realization that only the length of the string has an effect on the frequency of oscillation (Inhelder and Piaget, 1958). Each of the paper-and-pencil tests includes a task where a pendulum apparatus is illustrated, and the subject is asked to assess the control of variables in a double multiple choice format, giving both an answer and a reason for that answer. The paper-and-pencil tests were validated by comparing their correct and incorrect answers with those on Piagetian interview tasks, and they were found to have a statistically significant overlap.

Self-Regulation

The process through which an individual changes his or her mind is known as self-regulation. It is a process by which one compares current thinking patterns with information obtained from new experiences. If the new information is consistent with established mental understandings, no change in thinking is needed. When the new data conflicts with what is already known, an individual *may* change his or her mind, and if so, learning occurs.

Piaget said, "It is a process of [equilibration] in the sense—which has now been brought out so clearly by cybernetics—of self-regulation; that is, a series of active compensations on the part of the subject in response to external disturbances and an adjustment that is both retroactive (loop systems or feedbacks) and anticipatory, constituting a permanent system of compensations" (Piaget and Inhelder, 1969, p. 157).

Our observations indicate that a key difference between students who are successful in school and those who are not is the ability and/or desire to self-regulate. Students who self-regulate will consistently recognize when their current thinking patterns are not appropriate when given a source of new information, and then they will subsequently work to modify their thinking to be consistent with the new data. In essence, meaningful learning occurs whenever good students are presented with an opportunity and if they perceive that they have the time needed to learn.

Poor students generally choose to ignore the discrepancy between what they presently think and new, conflicting information. Instead of changing their thinking patterns, they will force new data into the wrong mental concept, ignore new data, or create a temporary new "this is how I'm supposed to answer a question in school" thinking pattern. This is not necessarily a conscious choice. In many cases, years of lack of practice in meaningful learning, copious reinforcement of "memorize and regurgitate equals success in school," and a social environment that discourages curiosity about learning leads to degradation of the internal driving force to self-regulate.

A key component in the definition of self-regulation is the word *self*. Meaningful learning is an internal process. First, an individual must pay attention to the new information presented. Second, they must compare that new information to how they presently think and recognize the conflict. Third, they must modify their thinking and test their new mental construct against additional data (Lawson and Wollman, 1975; Lawson, 1995; Lawson, 2003). All parts of this process must be done by an individual; a teacher cannot do it for them. Our job as an instructor is to provide *the opportunity* for students to self-regulate.

Piaget proposed that there are three factors that influence self-regulation: maturation, experience, and social transmission. Current evidence indicates that short-term memory capacity increases with age until it reaches its maximum of about seven bits somewhere in the mid-teen years. College instructors need not be concerned with maturation, but elementary and secondary school instructors need to be cognizant of the natural limitation of children's memories when constructing their curricula.

The role of experience is intuitively obvious. The more one interacts with the environment, the greater the opportunity to learn from those interactions. Interaction alone, of course, is necessary but not sufficient. Experience with the materials in our disciplines is the essence of why we consider laboratory to be an essential component of science coursework.

Social transmission has an essential role in the self-regulation process. Unfortunately, the environment for this type of interaction is not part of the traditional course structure. In order for people to change their minds, they must be exposed to conflicting information and alternate explanations for that information. Social interaction provides an environment where ideas can be debated, discussed, and argued. An individual's current way of thinking must be self-examined before it can be changed. It is interaction with peers and with teachers that leads to discussion and to meaningful learning.

Brain Physiology

There is mounting evidence, both psychological and physiological, that the human brain develops in a series of growth spurts interspersed with relatively dormant stages. Support for Piaget's developmental theory comes from both biological and psychometric brain research that shows growth spurts that coincide with the four stages of intellectual development. For example, Hudspeth and Pribram (1990) concluded that electroencephalogram data and other neurobiological data known at the time supported the conclusion that the brain develops in a series of growth spurts and plateaus in a manner consistent with Piaget's theories.

The work of J. N. Giedd, a neuroscientist at the National Institute for Mental Health, and collaborators, has profound implications for curriculum design in science courses (see, for example, Giedd et al., 1996; Giedd et al., 1999; Paus et al., 1999). Giedd's group has conducted *in vivo* cross-age studies of human brain development from age 4 to age 20 using magnetic resonance imaging. They have found that the gray matter of the brain continually thickens throughout childhood until it peaks as the child reaches sexual maturity. After the peak, the gray matter thins. They hypothesize that the neural connections that are not used are being eliminated throughout adolescence.

The implications of these studies are profound. First, the data provide physiological evidence in support of Piaget's theory that the potential for formal thinking ability begins with the onset of puberty, at about age 11 for girls and age 12 for boys. Second, the data indicate that if an adolescent is challenged to use their formal thinking ability (the gray matter is the location in the brain at which higher-order thinking occurs), those brain cells and their connections will survive. Contrarily, adolescents who are not in an environment where formal thinking is required will find that the development of formal thinking skills becomes much more difficult later in life. Giedd refers to this as the "use it or lose it" principle. It is imperative that middle school and high school science teachers challenge students to develop their formal thinking skills.

Psychometric measures designed to test prefrontal lobe activity have found notable regressions in inhibiting ability, planning ability, and mental capacity from the ages of 10 to 13 years. It is suggested that the maturation of the prefrontal lobes during adolescence is required for an individual to have the capacity to develop formal scientific reasoning abilities (Lawson and Kwon, 2000). Interestingly, a fifth brain growth spurt has been hypothesized at the age of 18, beyond the previous spurt at age 14 to 16 years that has been associated with the transition from Piaget's concrete to formal operational developmental stages. What this final brain growth represents is unclear, but current research suggests that it is linked with the ability to use the reasoning associated with formulating and testing alternative hypotheses about theoretical or abstract concepts (Lawson, et al., 2000b).

Applications of Theory

Conceptual Reorganization
Chi, Feltovich, and Glaser (1981) conducted a study that compared the categorization and representation of physics problems by experts and novices. The experts were advanced physics graduate students, and the novices were students who had just completed a semester of mechanics. The study found dramatic differences in the ways the subjects categorized physics problems. Novices tended to focus on literal surface features, whereas the experts used laws of physics as the basis for their categories.

A critical role of an instructor is in assisting students in making the transition from novice to expert. Consider the typical approach to stoichiometry in a first-term general chemistry course. Mass stoichiometry is typically the introduction to the topic. Solution stoichiometry follows in the next chapter. Textbook authors are pressured to keep chapters separate so that users can have the maximum flexibility with the book, so there is no mention of mass stoichiometry when solution stoichiometry is introduced. Gas stoichiometry and thermochemical stoichiometry follow later in the book, again with no formal links to earlier stoichiometry topics.

To help students make the transition from novice-level knowledge of these topics, which appear in different chapters and usually are tested on different exams, they need to be shown the relationship among the parts. All are stoichiometry problems—quantity relationships based on the ratios of reactant and product particles. Expert-level knowledge comes from moving to a higher level of generalization, and the curriculum design needs to incorporate a strategy for helping students reorganize their knowledge.

Vygotsky emphasized that conceptual reorganization is the critical psychological event that has to occur as a student moves from an intuitive understanding of a concept to a reflective understanding. We also believe that it is the critical difference between a novice level of knowledge in a discipline and an expert level. Vygotsky (1986) wrote, "If every concept is a generalization, then the relation between concepts is a relation of generality

(p. 197).... Th[e] position of a concept within the total system of concepts may be called its measure of generality (pp. 199–200).... the measure of generality determines ... all of the intellectual operations possible with a given concept (p. 201).... Furthermore, ... every new stage in the development of generalization is built on generalizations of the preceding level; the products of the intellectual activity of the earlier phases are not lost" (p. 202).

Of course, if students are not formal thinkers, they cannot see higher-order relationships among concepts, even if a course is designed to facilitate the development of expert-like knowledge. Thus, it is also imperative to have a curriculum design that encourages the development of formal thinking skills.

Behaviorism
Before constructivism became the theory of learning accepted by most science educators, behaviorism was the most influential theory in the United States. In the cognitive revolution, behavioristic theories were left behind, and a valuable body of literature reporting on a significant quantity of research soon became neglected. There are fundamental core facts that students should possess, application of those facts should be essentially a thoughtless process, and thus behaviorism does have a role, albeit small, in providing a theory base for educators. For example, given that human memory capacity is about seven bits (Miller, 1956), chunking separate data into a single datum is a crucial step in the process of complex thinking. Since we can think about approximately seven things at the same time, it pays to make each "thing" we think about as complex as possible. Chunking is primarily accomplished by repetition and practice.

Promoting Formal Thinking
Perhaps the most important work that demonstrates the potential for success of designing curricula that promote the development of formal thinking is that of the Cognitive Acceleration through Science Education (CASE) network, which is based in England. Adey and Shayer (1990) described their intervention in which one 60 to 80 minute science lesson was replaced by an intervention lesson every two weeks for two years in grades 7 and 8. The intervention lessons focused on formal thinking skills, such as controlling variables, proportional reasoning, combinatorial reasoning, and probabilistic reasoning. Interventions generally provided students with materials from which they might collect data, general descriptions to follow, and then asked how these materials were related to a particular outcome. Through interactions between the teacher and the student, students were challenged to describe how they knew their hypotheses were correct. In addition, these data-to-concept lessons were then related to specific examples from the curriculum, thus linking the target skills with content knowledge.

When compared to students from previous years using a standardized assessment of formal reasoning ability, Adey and Shayer (1990) found that their experimental groups moved from the 51st percentile on the pretest of formal reasoning skills to the 74th percentile on the posttest, while their control groups moved from the 52nd percentile to the 49th percentile, respectively. During this time, there was no significant difference in science achievement between the groups even though the experimental group replaced up to 25% of the original curriculum with these intervention lessons. These results provide strong evidence that teachers have the opportunity to enhance reasoning development when appropriate curricula and training are provided.

One year after this two-year intervention ended, Shayer and Adey (1992a) assessed students' gains in science content knowledge. For 12-year-old boys, the experimental groups showed a bimodal distribution of gains, which they classified as either low-gain or high-gain groups. They found an effect size of 0.33σ for the low group and an effect size of 2.24σ for the upper group relative to controls, where σ represents the standard deviation of the control group on the content assessment. In addition, the experimental groups of 11-year-old girls showed an effect size of 2.00σ for seven of the 30 girls in the group. Although an intervention would be effective at different ages for boys and girls, it is apparent that the intervention lessons focusing on formal reasoning skills enhance science achievement one year after the treatment ends.

Shayer and Adey (1992b) assessed student achievement two years after the end of the two-year intervention (age 16) using British National examination results. These results are coded A – G with an A being the highest mark possible. For boys' science achievement, 56.3% of the experimental group achieved a C or above, while only 12.5% of the control group achieved this score. For boys' mathematics achievement, 50% of the experimental group achieved a C or above, while only 14.3% of the control group achieved this score. The same trend was found in boys' English achievement: 39.3% of the experimental group achieved a C or above, while only 12.5% of the control group achieved this score. The girls' results were not significantly different for the groups in science and mathematics achievement (possibly due to the low numbers of girls continuing in science). However, for English achievement, 75.4% of the girls in the experimental group achieved a C or above, while only 70% of the control group achieved this score. Therefore, the intervention lessons in science enhanced not only achievement in science, but also achievement in areas as removed as English. This transfer provides evidence that targeting reasoning skills will enhance the general thinking skills of students.

The key aspects of the Shayer and Adey curricula were (a) a concrete introduction to the lesson, (b) a strategy for creating cognitive conflict, (c) a curriculum design that encourages the resolution of the disequilibrium created in part (b), with an emphasis on the development of one specific formal thinking skill in each activity, (d) metacognitive strategies to bring the target formal thinking skill into the consciousness of students, and (e) bridging activities that showed the utility of the target thinking skill in new contexts (Adey and Shayer, 1994).

Conclusions

At the beginning of the chapter, traditional teaching methods were subjected to criticism. It was stated that organizing curricula according to subject-matter-based principles and presenting the curricula with the traditional lectures-textbook-laboratory approach was inconsistent with how people learn. It is all too typical to hear science instructors blame students for their poor performance on course outcome measures. It is certainly true that, on average, students today work less at achieving intellectual growth while in college than the average student in the past (Sax et al., 2003), but the responsibility for improving curriculum design has been neglected by the majority of science instructors for half a century, so we must share the blame for lack of progress in student learning. This book provides a cornucopia of theory- and research-based techniques to help college chemistry students to effectively learn scientific thinking skills and chemistry content.

Acknowledgements

I am thankful to the editors of the book, Melanie Cooper, Tom Greenbowe, and Norb Pienta, for their insightful critique of an earlier draft of this chapter and for providing the opportunity to contribute. Additionally, John Deming and Kereen Monteyne, graduate students in our research group, have made countless contributions in the form of disequilibrating argumentation during our regularly scheduled discussions over the past few years, as well as through detailed critiques of an earlier draft of the chapter. I extend my gratitude to all.

Suggested Readings

Herron, J. D. (1996). *The chemistry classroom: Formulas for successful teaching*. Washington, DC: American Chemical Society.

Lawson, A. E. (2003). *The neurological basis of learning, development and discovery: Implications for science and mathematics instruction*. Dordrecht, The Netherlands: Kluwer Academic Publishers.

Vygotsky, L. S. (1986). *Thought and language*. Cambridge, MA: The MIT Press.

References

Adey, P. and Shayer, M. (1990). Accelerating the development of formal thinking in middle and high school students. *Journal of Research in Science Teaching*. 27(3), 267–285.

Adey, P., and Shayer, M. (1994). *Really raising standards: Cognitive intervention and academic achievement*. London: Routledge.

Ausubel, D. P. (1963). *The psychology of meaningful verbal learning*. New York: Grune and Stratton.

Ausubel, D. P, Novak, J. D., and Hanesian, H. (1978). *Educational psychology: A cognitive view* (2nd ed.)New York: Holt, Rinehart, and Winston.

Bodner, G. M. (1986). Constructivism: A theory of knowledge. *Journal of Chemical Education,* 63(10), 873–878.

Bodner, G., Klobuchar, M., and Geelan, D. (2001). The many forms of constructivism. *Journal of Chemical Education,* 78(8), 1107.

Bonham, S.W., Deardorff, D.L., and Beichner, R.J. (2003). Comparison of student performance using web and paper-based homework in college-level physics. *Journal of Research in Science Teaching,* 40(10), 1050–1071.

Bringuier, J. (1980). *Conversations with Jean Piaget.* Chicago: University of Chicago Press.

Bunce, D. M., Gabel, D. L., and Samuel, K. B. (1991). Enhancing chemistry problem-solving achievement using problem categorization. *Journal of Research in Science Teaching,* 28(6), 505–521.

Bunce, D. M., and Hutchinson, K. D. (1993). The use of the GALT (group assessment of logical thinking) as a predictor of academic success in college chemistry. *Journal of Chemical Education,* 70(3), 183–187.

Bybee, R. W., and Sund, R. B. (1990). *Piaget for educators* (2nd ed.) Prospect Heights, IL: Waveland Press.

Chi, M. T. H., Feltovich, P. J., and Glaser, R. (1981). Categorization and representation of physics problems by experts and novices. *Cognitive Science.,* 5, 121–152.

Chomsky, N. (1975). *Reflections on language.* New York: Pantheon.

Chomsky, N. (1980). *Rules and representations.* New York: Columbia University Press.

Clark, R. E. (1983). Reconsidering research on learning from media. *Review of Educational Research.* 53(4), 445–459.

Clark, R. E. (1994). Media will never influence learning. *Educational Technology Research and Development.* 42(2), 21–29.

Cole, R.S., and Todd, J. B. (2003). Effects of web-based multimedia homework with immediate rich feedback on student learning in general chemistry. *Journal of Chemical Education.* 80(11), 1338–1343.

Cracolice, M. S. and Roth, S. M. (1996). Keller's "old" personalized system of instruction: A "new" solution for today's college chemistry students. *The Chemical Educator.* 1(1). Available online at http://journals.springer-ny.com/chedr.

Deming, J. C., and Cracolice, M. S. (2004). Learning how to think. *The Science Teacher.* 71(3), 42–47.

Educational Policies Commission (1961). *The central purpose of American education.* Washington, DC: National Education Association.

Gabel, D. L., Sherwood, R. D., and Enochs, L. G. (1984). Problem-solving skills of high school chemistry students. *Journal of Research in Science Teaching.* 21(2), 221–233.

Giedd, J. N., Blumenthal, J., Jeffries, N. O., Castellanos, F. X., Liu, H., Zijdenbos, A., Paus, T., Evans, A. C., and Rapoport, J. L. (1999). Brain development during childhood and adolescence: A longitudinal MRI study. *Nature Neuroscience.* 2(10), 861–863.

Giedd, J. N., Snell, J. W., Lange, N., Rajapakse, J. C., Casey, B. J., Kozuch, P. L., Vaituzis, A. C., Vauss, Y. C., Hamburger, S. D., Kaysen, D., and Rapoport, J. L. (1996). Quantitative magnetic resonance imaging of human brain development: Ages 4–18. *Cerebral Cortex.* 6(4), 551–560.

Herron, J. D. (1975). Piaget for chemists. *Journal of Chemical Education.* 52(3), 146–150.

Herron, J. D. (1996). *The chemistry classroom: Formulas for successful teaching.* Washington, DC: American Chemical Society.

Herron, J. D., and Greenbowe, T. J. (1986). What can we do about Sue: A case study of competence. *Journal of Chemical Education.* 63(6), 528–531.

Hudspeth, W. J., and Pribram, K. H. (1990). Stages of brain and cognitive maturation. *Journal of Educational Psychology.* 82(4), 881–884.

Inhelder, B., and Piaget, J. (1958). *The growth of logical thinking* (A. Parsons and S. Milgram, Trans.) New York: Basic Books.

Keller, F. S. (1968). Goodbye, teacher…. *Journal of Applied Behavior Analysis.* 1, 79–89.

Kendall, J. (1935). *Smith's college chemistry.* New York: D. Appleton–Century Company.

Koumi, J. (1994). Media comparison and deployment: A practitioner's view. *British Journal of Educational Technology.* 25(1), 41–57.

Kozma, R. (1991). Learning with media. *Review of Educational Research.* 61(2), 179–211.

Kozma, R. (1994). A reply: Media and methods. *Educational Technology Research and Development.* 42(3), 11–14.

Lawson, A. E. (1978). The development and validation of a classroom test of formal reasoning. *Journal of Research in Science Teaching.* 15(1), 11–24.

Lawson, A. E. (1987). *Classroom test of scientific reasoning: Revised pencil-paper edition.* Tempe, AZ: Arizona State University.

Lawson, A. E. (1995). *Science teaching and the development of thinking.* Belmont, CA: Wadsworth.

Lawson, A. E. (2003). *The neurological basis of learning, development and discovery.* Norwell, MA: Kluwer Academic Publishers.

Lawson, A. E., Abraham, M. R., and Renner, J. W. (1989). *A theory of instruction: Using the learning cycle to teach science concepts and thinking skills.* Cincinnati, OH: National Association for Research in Science Teaching.

Lawson, A. E., Clark, B., Cramer-Meldrum, E., Falconer, K. A., Sequist, J. M., and Kwon, Y. –J. (2000). Development of scientific reasoning in college biology: Do two levels of general hypothesis-testing skills exist? *Journal of Research in Science Teaching.* 37(1), 81–101.

Lawson, A. E., Drake, N., Johnson, J., Kwon, Y. –J., Scarpone, C. (2000). How good are students at testing alternative explanations of unseen entities? *American Biology Teacher.* 62(4), 249–255.

Lawson, A. E., and Kwon, Y. –J. (2000). Linking brain growth with the development of scientific reasoning ability and conceptual change during adolescence. *Journal of Research in Science Teaching.* 37(1), 44–62.

Lawson, A. E., and Renner, J. W. (1975). Relationship of science subject matter and the developmental level of the learner. *Journal of Research in Science Teaching.* 12(4), 347–358.

Lawson, A. E., and Staver, J. R. (1989). Toward a solution of the learning paradox: Emergent properties and neurological principles of constructivism. *Instructional Science.* 18, 169–177.

Lawson, A. E., and Wollman, W. T. (1975). Physics problems and self-regulation. *The Physics Teacher.* 13(8), 470–475.

Marek, E. A., and Cavallo, A. M. L. (1997). *The learning cycle: Elementary school science and beyond* (revised ed.) Portsmouth, NH: Heinemann.

Masterton, W. L., and Slowinski, E. J. (1966). *Chemical Principles.* Philadelphia: Saunders.

Matthews, M. (1992). Constructivism and empiricism; an incomplete divorce. *Research in Science Education.* 22, 299–307.

Matthews, M. R. (1993). Constructivism and science education: Some epistemological problems. *Journal of Science Education and Technology.* 2(1), 359–370.

McKinnon, J. W., and Renner, J. W. (1971). Are colleges concerned with intellectual development? *American Journal of Physics.* 39(9), 1047–1052.

Miller, G. A. (1956). The magical number seven, plus or minus two: Some limits on our capacity for processing information. *Psychological Review. 63,* 81–97.

Nola, R. (1997). Constructivism in science and science education: A philosophical critique. *Science and Education.* 6, 55–83.

Novak, J.D., and Gowin, D.B. (1984). *Learning how to learn.* New York: Cambridge University Press.

Nurrenbern, S. C., and Pickering M. (1987). Concept learning versus problem solving: Is there a difference? *Journal of Chemical Education.* 64(6), 508–510.
O'Donohue, W., and Kitchener, R. (1998). *Handbook of behaviorism.* San Diego: Academic Press.

Parkin, A.J. (2000). *Essential cognitive psychology.* Philadelphia: Taylor and Francis.

Paus, T., Zijdenbos, A., Worsley, K., Collins, D. L., Blumenthal, J., Giedd, J. N., Rapoport, J. L., and Evans, A. C. (1999). Structural maturation of neural pathways in children and adolescents: In vivo study. *Science.* 283, 1908–1911.

Piaget, J. (1950). *The psychology of intelligence.* London: Routledge and Kegan Paul.

Piaget, J. (1962). *Play, dreams, and imitation in childhood.* New York: Norton.

Piaget, J. (1967). *Six psychological studies.* New York: Random House.

Piaget, J. (1977a). *The development of thought: Equilibrium of cognitive structures.* New York: Viking.

Piaget, J. (1977b). *The grasp of consciousness.* London: Routledge and Kegan Paul.

Piaget, J. (1985). *The equilibrium of cognitive structures: The central problem of intellectual development* (T. Brown and K. L. Thampy, Trans.) Chicago: University of Chicago Press.

Piaget, J. and Inhelder, B. (1969). *The Psychology of the Child.* New York: Basic Books.

Piattelli–Palmerini, M. (Ed.) (1980). *Language and learning: The debate between Jean Piaget and Noam Chomsky.* Cambridge, MA: Harvard University Press.

Quartz, S. R. (1993). Neural networks, nativism, and the plausibility of constructivism. *Cognition.* 48, 223–242.

Resnick, L. B. (1980). The role of invention in the development of mathematical competence. In R. H. Kluwe and H. Spada (Eds.), *Developmental models of thinking.* New York: Academic Press.

Roadrangka, V., Yeany, R. H., and Padilla, M. J. (1982). GALT: Group test of logical thinking. Athens: University of Georgia.

Sanderson, R. T. (1963). *Principles of chemistry*. New York: Wiley.

Sax, L. J., Astin, A. W., Lindholm, J. A., Korn, W. S., Saenz, V. B., and Mahoney, K. M. (2003). *The American freshman: National norms for fall 2003*. Los Angeles: Higher Education Research Institute.

Scerri, E. R. (2003). Philosophical confusion in chemical education research. *Journal of Chemical Education.* 80(5), 468–474.

Shayer, M. and Adey, P.S. (1992a). Accelerating the development of formal thinking in middle and high school students II: Postproject effects on science achievement. *Journal of Research in Science Teaching.* 29(1) 81–92.

Shayer, M. and Adey, P.S. (1992b). Accelerating the development of formal thinking in middle and high school students III: Testing the permanency of effects. *Journal of Research in Science Teaching.* 29(10) 1101–1115.

Sienko, M., and Plane, R.A. (1966). *Chemistry: Principles and properties*. New York: McGraw-Hill.

Skinner, B.F. (1974). *About behaviorism*. New York: Alfred A. Knopf.

Sternberg, R.J. (2002). *Cognitive psychology* (3rd ed.) Belmont, CA: Wadsworth.
Szabo, A., and Hastings, N. (2000). Using IT in the undergraduate classroom: Should we replace the blackboard with PowerPoint? *Computers and Education.* 35(3), 175–187.

Tobin, K. G., and Capie, W. (1980a, April). *The Development and Validation of a Group Test of Logical Thinking*. Paper presented at the annual meeting of the American Educational Research Association, Boston, MA.

Tobin, K. G., and Capie, W. (1980b, April). *The Test of Logical Thinking: Development and Applications*. Paper presented at the annual meeting of the National Association for Research in Science Teaching, Boston, MA.

Valanides, N. (1999). Formal reasoning performance of higher secondary school students: Theoretical and educational implications. *European Journal of Psychology of Education.* 14(1), 109–127.

von Glassersfeld, E. (1981). Concepts of adaptation and viability in a radical constructivist theory of knowledge. In Sigel, I., Brodzinsky, D., and Golinkiff, R. M. (Eds.), *New directions in Piagetian theory and practice*. Hillsdale, NJ: Lawrence Erlbaum Associates.

von Glassersfeld, E. (1995). *Radical constructivism: A way of knowing and learning*. Washington, DC: Falmer.

Vygotsky, L. S. (1986). *Thought and language*. Cambridge, MA: The MIT Press.

Wittrock, M. C. (1974). Learning as a generative process. *Educational Psychologist.* 11, 87–95.

3

All Students are Not Created Equal:
Learning Styles in the Chemistry Classroom

Stacey Lowery Bretz, Ph.D.
Department of Chemistry
Youngstown State University

Abstract

"Learning styles" is part of the pedagogical lexicon, but no one can point to a precise definition. "Learning styles" is a construct that encompasses a wide variety of strategies and preferences expressed by students for how they like to learn, and how they don't like to learn; for what they believe helps them learn, and for what they believe interferes with their learning. A variety of schema, i.e., categorization systems, exist to describe students and these strategies and preferences. Yet their value lies not in identifying which one is best or most powerful, but in using them as a vehicle to help students reflect upon their strengths and weaknesses as a learner. Therefore, teachers have a responsibility to challenge their students to continue to grow cognitively and to provide mechanisms to support that growth. This chapter summarizes the literature regarding a variety of learning styles schema with a discussion of their implications for teaching in the chemistry classroom and laboratory.

Biography

Stacey Lowery Bretz is a Professor of Chemistry at Youngstown State University. Dr. Bretz earned her B.A. and M.S. in chemistry at Cornell University and Penn State University, respectively, before returning to Cornell to earn a Ph.D. in chemistry education research (CER). After completing a post-doctorate in CER at Berkeley, Dr. Bretz spent five years on the faculty at University of Michigan–Dearborn working on curriculum reform in general chemistry. Dr. Bretz's current research focuses upon the assessment of student learning, the professional development of chemistry teachers, and the application of cognitive science theories and qualitative methodologies to CER. Dr. Bretz will chair the Gordon Conference on CER in 2005 and currently serves on the Board of Trustees for the American Chemical Society Division of Chemical Education Examinations Institute. She is a recipient of the Distinguished Professor of Teaching Award at YSU, the Northeast Ohio Council of Higher Education Excellence in Education Award, and has been selected by *Ohio* Magazine as one of the Top 100 Faculty Members in Higher Education.

Introduction

Consider the following scenario: Professor Jones, an organic chemist, in an effort to improve her teaching, makes the following observations about three of her better students:

- Suzannah takes meticulous notes as well as tape-records all lectures. She diligently completes homework, using the study guide to look up answers. Suzannah performs well on multiple choice questions on the exams, but struggles with open-ended questions, e.g., synthesis from mono-functional compounds and multi-step mechanisms.

- Joe, on the other hand, takes almost no notes during lecture, yet listens very attentively. He rarely asks questions in lecture, yet regularly comes to office hours to discuss his understanding or any points of confusion. He often uses molecular models during office hours to make his point.

- Mikaela, who takes notes and clearly pays close attention, regularly talks to her classmates during lecture, apparently not to socialize, but to discuss the concept at hand. (Whenever Professor Jones asks if she has a question, Mikaela never becomes quiet as do students who are embarrassed to be caught chatting; she always has a substantive question, often asking for real-world examples or applications.) Mikaela regularly asks Professor Jones how the topic at hand is related to similar ideas in her microbiology class.

After observing these three students, Professor Jones is somewhat confused. How can these three students, all with very different strategies when it comes to learning in lecture, all do equally well on her exams? What can she do to improve the learning for all of the students in her organic chemistry class when it appears that success can be attributed to a collection of such very different learning styles?

What is a learning style?

Although no consensus currently exists on a singular definition of learning style, at its core is the idea of a skill set which students use to perceive and process new information. According to Richard Felder, Professor of Chemical Engineering at North Carolina State University, students have different learning styles because "they preferentially focus on different types of information, tend to operate on perceived information in different ways, and achieve understanding at different rates" (Felder, 1993). Consider our organic chemistry students, for example. Note-taking is important to Suzannah, and she double-checks her notes by listening to her tape recording; by contrast, Mikaela takes notes, but prefers to gauge her understanding by checking with her classmates. And then there is Joe, who prefers to listen intently and take few, if any, notes.

Synonyms for "learning styles" abound in the literature: learning strategies, learning preferences, learning modalities, learning orientations, and learning differences. Regardless of the name, the construct is multi-dimensional and a variety of schema, i.e., systems to categorize students, have been devised and tested in order to explain the multiplicity of student learning outcomes. While each scheme presents a differing perspective, with its own system of nomenclature, there are commonalities. Learning styles can include the physical conditions of the environment a student finds conducive to studying, concentration, and learning. Sensory modes of perception, the degree of social interaction, and cognitive processes are additional dimensions invoked to explain learning styles.

This chapter aims to review the theory and practice of learning styles as they relate to college chemistry teaching, specifically to (1) introduce a variety of learning styles, (2) discuss this research in the context of the chemistry classroom and laboratory, and (3) ultimately, cultivate a recognition and appreciation of learning as a complex, variable, and dynamic phenomenon. Chemists like Professor Jones in the introduction will recognize their own students (if not themselves) in the following pages, and in doing so, develop a deeper understanding of the diversity of ways that students learn chemistry.

Perceiving and Processing Information

From Carl Jung's seminal work (Jung, 1971) on sensing and intuition as primary modes by which people perceive the world around them, to the familiar "left brain/right brain" dichotomy of Herrmann's Brain Dominance theory (Herrmann, 1990; Springer and Deutsch, 1985), the concept of learning styles has not been just an academic exercise, but has crossed over to public consumption. Howard Gardner's *Frames of Mind* and his framework of "multiple intelligences" has once again stirred public debate over how students learn (Gardner 1983, 1993).

Four different schema for learning styles, each with particular relevance to learning chemistry, are listed on the next page:

- VARK

- Myers-Brigg Personality Type Indicator

- Kolb's Experiential Learning Model

- Felder-Silverman Index of Learning Style

After an introduction to the nomenclature for each of these schemes, we turn to the challenges of learning in a chemistry classroom, followed by a discussion of the relevance of learning styles to these challenges.

VARK

Certainly students not only must be able to take in information, but they also must be capable of communicating what they know in order to demonstrate that learning has taken place. One part of a student's learning style is a preference for the mode through which they take in information as well as the mode by which they espouse their knowledge to convey their learning. VARK, which stands for Visual–Aural–Read/Write–Kinesthetic, offers learners the opportunity to construct a profile of their preferences (Fleming, 2001).

Students with a strong preference for visual (V) strategies will benefit from graphs, symbols, or concept maps, while lectures and discussions are well-suited to those with aural (A) strategies, such as Suzannah in the introduction to this chapter. Students with reading/writing (R) preferences opt for strategies including lists, outlines, and definitions. Lastly, kinesthetic (K) preferences for learning are typified by engaging students in field trips, laboratories, and role plays—remember Joe's use of molecular models to explain his ideas.

Although sensory perception is just one of the research-identified components of a learning style, Fleming argues (2001) that VARK offers particular promise because students can learn to manipulate their sensory inputs and outputs. For example, Professor Jones might encourage all of her students to try Joe's strategy of listening to lectures (an aural preference), rather than trying to transcribe them through meticulous note-taking. She could take a few moments to discuss with the entire class their experience of shifting to listening.

In fact, both teachers and students need to guard against viewing the labels of VARK as permanent. Teachers should caution students that, "*once a visual learner* does not mean *always a visual learner* (or kinesthetic or auditory or …)":

> "*Learning styles ought not be considered immutable and hierarchic (as cognitive science implies), but rather ... as 'learning preferences' or 'learning strategies' into which the student has fallen because of having been rewarded (or self-reinforced) through his or her years of schooling. As more and more high school subjects are made elective, the typical student is not forced to stretch beyond a certain 'comfort zone' of learning preferences.*" (Tobias, 2002).

Being an effective chemistry teacher requires "stretching" students, as Tobias puts it, to broaden their repertoire of learning strategies.

Myers–Brigg Personality Type Indicator

The Myers–Brigg inventory (Lawrence, 1997), while used to facilitate hiring practices and compatibility of team members in the workplace, has become a popular staple in the classroom to introduce students themselves to learning about their preferences for learning. MBI consists of four dimensions:

- introversion (I) / extroversion (E)

- sensing (S) / intuition (N)

- thinking (T) / feeling (F)

- perceiving (P) / judging (J)

The first two of these dimensions, I/E and S/N, measure differences as to how students take in and process information, i.e., how they perceive the world around them. The latter two, T/F and P/J, then, account for differences in how students behave in the face of their perceptions, i.e., what actions do they take as a result of their inputs.

Therefore, a student could be classified as INTJ. Such a student would prefer to focus on ideas and concepts (I) and to embrace abstractions and generate possibilities (N), to judge new information objectively and analytically (T), and all the while prefer control and decisiveness (J). For example, a chemistry student with these Myers–Briggs characteristics would be comfortable with invoking abstract concepts (e.g., the use of shielding to explain periodic trends) or mastering the conventions of molecular orbital theory.

By contrast, an ESFP learner would be driven by his need to connect to the world around him (E), to perceive the world through sensory experiences (S), to focus on the personal and subjective in learning through empathy (F), and to place a value on adapting and understanding new situations (P). Felder's analysis of how the T/F and P/J dimensions play out in a conceptually abstract, mathematically rich subject like engineering seem equally valid for chemistry:

> "*Thinkers like logically organized presentations of course material and feedback related to their work; feelers like instructors who establish a personal rapport with them and feedback that shows appreciation of their efforts. Judgers like well-structured instruction with clearly defined assignments, goals, and milestones; perceivers like to have choice and flexibility in their assignments and dislike having to observe rigid timelines.*" (Felder, 2002).

Think about the chemistry classrooms in which you have been a student or a teacher. Was the course structured to fit the preferences of thinkers? Feelers? Judgers? Perceivers? Chemistry faculty typically give highly detailed and organized lectures. Chemists routinely develop syllabi that expect conformity: problem sets of required problems with firm deadlines. Chemists assess student learning primarily through exams, often providing little formative feedback, but more often just a judgment of "correct" or "wrong," accompanied by a proportional loss of points. Clearly, the predominant culture of chemistry classrooms strongly favors some of the MBI dimensions while minimizing, if not outright ignoring, others.

Kolb's Experiential Learning Model

While Fleming's VARK modalities focus upon the role our senses play in learning, David Kolb emphasizes the importance of experience in his model to explain differences in learning (Kolb, 1984). These differences stem from the variability in how people perceive their experiences and the processes by which they make meaning of those experiences, as well. The perception dimension of this variability encompasses learners who require concrete, sensory experiences for learning as well as those who thrive on abstraction and symbolism. The processing dimension, which represents how learners transform or make meaning of their experiences, recognizes those who learn by actively experimenting or manipulating the situation as well as those who prefer to learn though quiet reflection. Figure 1, adapted from Towns (2001), shows the intersection of the active/reflective continuum and the concrete/sensory continuum, resulting in the creation of Kolb's four learning styles: divergers, assimilators, convergers, and accommodators.

Central to Kolb's model is the recognition that students in different quadrants will focus upon different aspects of understanding a given concept. Divergers will ask *Why is this important to know?* while assimilators want to know *What is the concept?* Convergers will ask *How is this concept applied?* while accommodators wonder *What are the possibilities of this concept?* Consider, for example, the concept of chemical bonding. Divergers will be interested in real-life examples that convey the ideas behind bond type, bond length, bond strength, etc.; assimilators, on the other hand, will seek out definitions and rules for classifying the kinds of bonds. Convergers will be interested in testing out the idea of bonding in other disciplines (like Mikaela in the introduction and her microbiology class). Accommodators will want to explore the importance of bonding to explain related concepts and experimental observations, such as thermochemistry or reaction kinetics.

Felder–Silverman Index of Learning Styles

Felder has written (1993) that a "student's learning style may be defined in part by the answers to five questions:

1. What type of information does the student preferentially perceive: *sensory*—sights, sounds, physical sensations—or *intuitive*—memories, ideas, insights?
2. Through which modality is sensory information most effectively perceived: *visual*—pictures, diagrams, graphs, demonstrations—or *verbal*—sounds, written and spoken words and formulas?
3. With which organization of information is the student most comfortable: *inductive*—facts and observations are given, underlying principles are inferred—or *deductive*—principles are given, consequences and applications are deduced?
4. How does the student prefer to process information: *actively*—through engagement in physical activity or discussion—or *reflectively*—through introspection?
5. How does the student progress toward understanding: *sequentially*—in a logical progression of small incremental steps—or *globally*—in large jumps, holistically?"

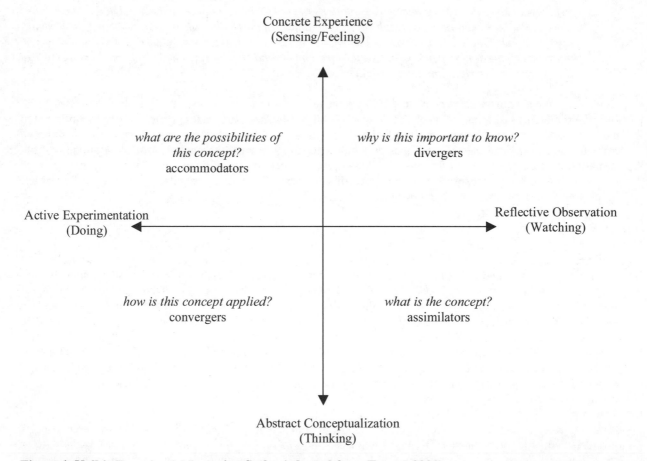

Figure 1. Kolb's Experiential Learning Styles (adapted from Towns, 2001).

Analogous to classifying students per the Myers–Brigg model, a student under Felder–Silverman would be labeled with regard to each of the five categories described above. For example, a student's learning style may be described as sensory, visual, inductive, reflective, and sequential. Consider once again Professor Jones' students in the introduction: Mikaela's desire to discuss ideas with her peers typifies an active learning style whereas Joe's quiet, attentive listening is reflective.

In contrast to Fleming's VARK model, however, Felder and Silverman have argued that *verbal* is a better descriptor than *auditory*, particularly in dealing with the processing of written words. As written words are read, not pronounced aloud, they are clearly not auditory. Furthermore, research on second language acquisition supports the claim that written words are processed in a manner analogous to the spoken word (Felder and Henriques, 1988).

Recently, Felder dropped the inductive/deductive dimension from the model for fear that "instructors [would] give our instrument to students, find that the students prefer deductive presentation, and use that result to justify continuing to use the traditional deductive instructional paradigm in their courses and curricula" (Felder and Soloman, 2004). An honest self-assessment by chemistry faculty reveals why Felder expects chemistry students to prefer deductive teaching. It matches the preferred teaching style of chemistry faculty. We prefer to share our knowledge of chemistry in an efficient manner, meaning we talk as students listen. The body of knowledge that is chemistry is continually growing, forcing us to make choices about what to present and how to present it. Efficiency demands that we cover material quickly. All of which leads to the following scenario being played out over and over in lecture halls:

> *Professor Smith begins the next chapter by defining important terms and concepts, offering perhaps a carefully chosen example to support the definition. Then, to show students how these concepts are connected, he says, "It turns out that..." and writes a mathematical equation on the board, in turn defining each symbol and term. Next, he works a problem to show the utility of the equation. If time remains, he comments upon the application of this equation to a real-world example. Class dismissed. Next class, the routine begins again.*

Look closely at the verbs in this description: *define, offer, show, write, works, comments*. Whose actions are these? Clearly, they are the teacher's. What actions does a student take in this scenario? What possible learning strategies would be successful? It is not unreasonable to describe the student's responsibility as "keeping up with" the professor as he moves from offering definitions and principles to demonstrating utility and application.

While such an arrangement can indeed by efficient, it also raises serious questions. What impressions do students form about how chemistry knowledge is constructed? It is not surprising that students perceive these facts and principles to be truths known since time began—not as the constructions and models forged by human experimentation, fraught with many false starts, blind alleys, and oversimplifications.

In short, deductive teaching makes sense to faculty because we already understand the chemistry. Yet, when most faculty stop to reflect upon how they learn new information, e.g., in the research laboratory, we realize that learning takes place in an inductive fashion—we move from gathering observations and data to forming generalizations.

The diversity of learning styles described in this chapter creates a dilemma for faculty who prefer deductive teaching: how can you restructure the classroom to help students learn inductively? How can learning styles help a chemist be a more effective teacher?

How can learning styles help a chemist be a more effective teacher?

According to Porter and Brophy (1988), effective teaching is both a *process*-based and *content*-based enterprise (italics original). Effective teachers

- know the subject matter they want their students to learn.

- know the misconceptions their students bring to the classroom that will interfere with their learning of that subject matter.

- provide students with metacognitive strategies to use in regulating and enhancing their learning.

- provide students with structured opportunities to exercise and practice independent learning strategies.

By now, you, the reader, may very well be wondering, *Which learning styles and strategies should I legitimize and/or promote? Is one scheme better than the others?* VARK and Kolb have four styles each, while Myers–Brigg and Felder–Silverman have 2^4 or 16 different styles! The thought of measuring and accommodating up to 16 different styles in one classroom may leave you more paralyzed than inspired to be particularly effective. And yet, to focus exclusively on this aspect of learning styles is to miss an opportunity to significantly impact your students' learning. Sober yourself by remembering these two points:

1. Teachers are not expected to, nor capable of, accommodating each and every learning style within each and every classroom activity or assessment. Too much responsibility is placed on the teacher; not enough on the student (Tobias, 2002).

2. Students can and do use more than one learning style. No student is 100% kinesthetic or 100% diverger. In fact, both the Myers–Brigg Inventory and the Felder–Silverman model embrace this diversity with the multidimensional nature of their schema.

Metacognition and Meaningful Learning

The issue then is not how, or even if, teachers can accommodate the entire spectrum of learning styles. For inevitably, students can and will embrace particular learning strategies, and when they do, teachers need to understand that student preferences are a direct extension of what that student believes it means "to know" something. And therein lies the true power of the construct of learning styles—not which one is right or most powerful, but rather that they offer teachers a gateway to nurturing their students' metacognitive abilities. Yet, paradoxically, the work of metacognition lies in the province of the student; the teacher cannot impose such development upon a student.

Flavell (1985) has defined metacognition as "knowledge concerning one's own cognitive processes." Rickey and Stacy articulated the connection between metacognition and learning styles by drawing distinctions between general strategies which cross disciplinary boundaries and strategies specifically suited to a particular discipline (in our case, chemistry):

> *"An example of a general thinking strategy is evaluating new ideas by comparing them to other things you know about to see if the ideas fit together and make sense to you. A strategy more specific to the domain of chemistry is making connections between macroscopic observations and molecular-level explanations."* (Rickey and Stacy, 2000).

So as a student learns more about her preferences and her well-developed learning strategies, she can begin to forge connections between those general strategies and her deepening knowledge of chemistry. She is developing the self-knowledge of metacognition. Forging such connections, however, requires more than just recognition of one's preferred learning strategies, be they consistent with Kolb, VARK, or some other model. Forging such connections requires students to commit to meaningful learning.

The idea of meaningful learning stands in direct contrast to rote learning, in which the new material is not connected in any substantive manner to existing knowledge, but is merely memorized. Ausubel (1963, 1968) speaks of meaningful learning as the processes by which new knowledge is incorporated into the learner's mind. In order for students to escape memorization, they must choose to make connections between what they *already know* and what they *need to know* (Novak and Gowin, 1984; Bretz, 2001). Helping students to make these connections, that is, *to learn how to learn*, is the central tenet of Ausubel and Novak's construct of meaningful learning (Ausubel 1968; Ausubel, Novak, and Hanesian, 1978).

Learning in the Chemistry Laboratory

Lecture is not the only place that chemistry students can be "'stretched" out of their comfort zone for meaningful learning; the laboratory offers many possibilities for helping students learn how to learn chemistry. At one end of the continuum, there are those students who struggle in the chemistry laboratory; they seem to have little understanding of what to do or when to do it, let alone why. Their strategy for lab (it would be an exaggeration to call it a "'learning" strategy) is one of survival—they often copy their classmates' procedures, or just partner with a peer who does it all. At the other end of the spectrum are students who not only carry out the procedures flawlessly, but also seem to understand the purpose behind each step of the experiment. And in between is a vast space of possibilities for students to move from rote mechanics to meaningful understanding. This middle group of students can often obtain products of high quality in an excellent yield, yet they seem very unsure of themselves. They frequently ask for reassurance about procedures (*"Is this what I'm supposed to do next?"*) or the consequences of poor performance (*"The directions say to measure out 2.00g. I have 2.05g. Is that going to mess up my experiment?"*). There seems to be a disconnect between the ability to do the chemistry on the level of beakers and grams and the ability to understand the chemistry happening on the level of atoms and molecules. (These are the students who never seem to know which layer they want during extractions!)

How do we explain this gulf of confusion between what students are supposed learn about chemistry in lecture and what they seem to be unable to put into action in the chemistry laboratory? Robertson (1984) interviewed college students who had been videotaped during their laboratory session in an introductory biology course. By showing the videotape to these students during an interview shortly after the laboratory was completed, Robertson probed for the nature of the relationship among students' thoughts, feelings, and actions during the laboratory. She classified students' learning strategies in the laboratory as *think as little as possible* (actions in the laboratory accompanied by no conceptual understanding), *think procedurally* (adept execution of the laboratory experiment, but little cognition), and *think meaningfully* (meaningful integration of actions with the concepts in addition to adept execution of the procedures). Robertson's study forces chemistry faculty to confront an unpleasant truth: students can and do leave our courses having completed lab experiments without a rich understanding of the concepts underlying the measurements they make. They are able to do so because of the disconnect between students' learning styles and how we measure learning.

Additional work in the field of cognitive development would suggest that the same learning strategies which students employ during lab are also employed during lecture. Edmondson's investigation of the connection between college students' learning strategies and science concepts argues that:

> *"If, for example, science is thought of as a body of proven facts, a person might study and memorize the facts, and might think that he or she can absolutely prove things in science. If a person thinks of science as an ongoing process of concept development, he or she might [learn] concepts and their variants."* (Edmondson, 1989 as quoted in Waterman, 1982, p. 5)

Edmondson identified that students who were dualistic receivers of knowledge (Perry, 1968; Finster, 1989, 1991) tended to use rote learning strategies exclusively, whereas those students whose conception of knowledge was consistent with constructivism were likely to employ meaningful learning strategies. Interestingly, Edmondson also observed a third group of students who maintained dual, but separate, systems of learning strategies as a result of their compartmentalization of knowledge across subject areas. These students, while skilled at adapting their learning strategies to the demands of the class or task at hand, showed little ability to integrate their knowledge across courses or subject areas (Edmondson and Novak, 1993). This brings us to ask, *What factors cue a student to adapt his learning strategy?*

Deep and Surface Learning Strategies

To answer this question, we turn to the work of Entwistle and Ramsden (1983) who developed a questionnaire to characterize a student's approach to studying as either *deep* or *surface*. Students with a "surface" approach to learning are mainly concerned with memorizing whatever information they considered to be important—a consideration usually informed by the exam questions and assessment measures encountered in the classroom. Students using the "deep" approach are concerned with understanding the meaning of content in light of their prior knowledge and experiences.

In a study attempting to shift college biology students from using a rote, surface approach to assuming a more meaningful learning orientation, Donn found (1990) that meaningfully-oriented students responded to novel problems by reflecting questions back to the interviewer and by formulating relationships between ideas. By contrast, learners with a rote orientation were inclined to respond with verbatim definitions and were often stymied by questions attempting to probe beyond their memorization of facts and definitions. Perhaps more astounding, though, was the result that despite the introduction of a metacognitive tool known as the Vee diagram (Gowin, 1981), which was designed to explicitly facilitate a student's growth from rote memorization to construction of knowledge,

> "*there was a shift toward a surface approach for both males and females ... the interpretation offered is that the underlying epistemology of the discipline of science as supported by the grading and teaching practices exerts a more powerful influence than an isolated attempt to alter students' basic views on the construction of knowledge.*" (Donn, 1990, abstract)

That is to say, the teacher's choice about what to assess and how to assess is actually more influential upon student learning strategies than the students' own preferences.

Consequences of Learning Styles for Chemistry

Donn's observation that learning strategies do not operate in isolation from the culture of the classroom, specifically the teaching and assessment practices of a discipline, is right on target. In 1990, Sheila Tobias published a study entitled *They're Not Dumb, They're Different*, in which she examined the teaching and assessment practices of introductory university chemistry (and physics) courses in order to better understand what made science hard for those students she identified as members of the "second tier:"

> "*The second tier is a loose hypothetical construct, which includes a variety of types of students not pursuing science in college for a variety of reasons. They may have different learning styles, different expectations, different degrees of discipline, different 'kinds of minds' from students who traditionally like and do well at science.*" (Tobias, 1990, p. 14; italics own)

Tobias' study recruited stand-ins for the second tier, six graduate students and one professor from nonscience fields, to "seriously audit" introductory physics and chemistry courses with the purpose of monitoring their own experiences as well as the classroom "culture" of beginning science courses for indicators of what makes science hard. The experiences of these students bear out Tobias' admonition that nonscience students cannot be labeled as such for any singular reason. They choose majors outside the sciences for a variety of reasons, including preferences for different learning styles, expectations, and "kinds of minds." Consider the inflexibility of science for one student's preferred method of learning:

> "*I process information in a different way than it is taught or utilized in science courses. I learn to understand by putting [concepts] into my own language, not by memorizing and spitting out the words as I receive them. [Also] the humanities and social sciences seem to be patient with people like me ... Science seems to hurry off before I get too close. It avoids my attempts to touch or shape it.*" (Tobias, 1990, p. 54)

Being able to "touch or shape" the subject matter is an important expectation for good nonscience students. They point to opportunities in the humanities and social sciences allowing for their input through discussion of the subject, even at the introductory level. Science is presented to them as a finished product, with no opportunities for discussion unless one perseveres through many more years of coursework to get to the "cutting-edge" of science. Consider the journal entry of one of the auditors regarding her experience in a college chemistry course:

> "*Chemistry is a very hard and fast science. Facts are facts and at the introductory level there is little debate about what is presented to students. Basically, there is a body of information that has to pass from professor to student and there is no room for interpretation or creative thought.*" (Tobias, 1990, p. 52)

This student's complaint echoes the findings of Waterman (1982) and Edmondson (1989) who concluded that one consequence for students who view science as a body of absolutely proven facts is to use rote learning strategies exclusively.

Tobias' study lends support to the idea that some nonscience students expect and hope for their science classes to accommodate meaningful learning strategies they find successful in their humanities and social sciences studies. The observations and experiences of Tobias' auditors in the college science classroom identify many problems with the manner in which science is taught, learned, and assessed. These critiques are echoed in the AAAS report, *Science for All Americans*:

> "... *science textbooks and methods of instruction ... emphasize the learning of answers more than the exploration of questions, memory at the expense of critical thought, bits and pieces of information instead of understandings in context, recitation over argument, reading in lieu of doing. They fail to encourage students to work together, to share ideas and information freely with each other, or to use modern instruments to extend their intellectual capabilities.*" (AAAS, 1989, p. 14)

In spite of the shortcomings of chemistry as it is experienced by many students, including the "second tier," the practice of science as an intellectual endeavor requires students with strengths across a variety of learning styles. As Felder writes,

> "*Students whose learning styles fall in any of the [ILS] categories have the potential to be excellent scientists ... observant and methodical sensors make good experimentalists ... insightful and imaginative intuitors make good theoreticians ... active learners are adept at administration and team-oriented project work ... reflective learners do well at individual research and design ... sequential learners are good analysts ... global learners are good synthesizers able to solve problems from multi-disciplinary perspectives.*" (Felder, 1993)

How, then can teachers communicate this message to their students? How can we help them to experience science, to learn science, to "stretch" their zone of cognitive comfort as Tobias suggests?

Effective Teaching: Teachers' Responsibilities for Learning Styles

We teach as we have been taught. How were we, teachers of chemistry, taught chemistry? Lectures, chock-full of deductive reasoning. While the veracity of this maxim is slowly eroding in chemistry, it still holds true for a significant number of chemists.

We teach as we learn. How do we, teachers of chemistry, learn chemistry? Fleming's research (2001) regarding preferred modes of sensory input and output shows that faculty have a strong preference for read/write strategies. Not surprisingly, most of our assessment of student learning comes in written form.

Our students do not necessarily process information as we do. Our students are much stronger than we in kinesthetic modes. How many of us can truly say we structure our assessments to benefit this learning style?

The triangulation of the italicized statements beginning each of the three paragraphs above creates quite a challenge for teachers of chemistry. Being effective means successfully performing a "balancing act." Effective teachers need to balance their comfort zone of lectures against their responsibility to provide a multitude of ways for students to interact with the content. Effective teachers need to balance their comfort zone of written test questions and answers against their responsibility to construct and value assessments that will permit their students to show abilities across the diversity of modalities described in this chapter. Effective teachers will make good choices for their students guided by this wisdom from the *National Science Education Standards*:

> "*Learning science is an active process. Learning science is something students do, not something done to them.*" (National Research Council, 1996).

The remaining chapters in this book are overflowing with research-based evidence to support the good choices of effective teachers. Michael Abraham discusses the role of guided inquiry in laboratory learning, a strategy well connected to kinesthetic learners and convergers. Mary Kay Orgill and George Bodner discuss the role of analogies in learning chemistry, Mary Nakhleh presents the research findings on concept mapping, and Roy Tasker writes about the importance of visualization in learning chemistry. Melanie Cooper, Pratibha Varma-Nelson, and Brian Coppola discuss the significance of cooperative learning and team learning in encouraging dialogue about chemistry and its abstract concepts. All of these chapters provide concrete methods by which teachers can fulfill their responsibility to meet the diversity of learning styles within their classroom.

The Search for Shared Meaning

Gowin has written (1981) that to educate is to change the meaning of experience and that meaning is constructed through shared experience. The challenge for teachers and students of chemistry, then, is to achieve shared meaning. Students enter the chemistry classroom and laboratory with varied educational and real-life experiences. The sum of that student's prior knowledge and experiences with science, including their learning styles, will influence the integration of new knowledge. Teachers have a responsibility to help the student strive to employ meaningful learning strategies, i.e., to help the student distinguish the similarities, differences, and connections between the new and the existing knowledge.

In the epigraph to *Educational Psychology: A Cognitive View*, Ausubel (1968) writes:

> "*The most important single factor influencing learning is what the learner already knows. Ascertain this and teach him accordingly.*"

Certainly, effective teachers need to know their chemistry and what misconceptions students harbor regarding the content. Students' preferred learning styles are an important factor to consider when accounting for what "the learner already knows."

Suggested Readings

Felder, R. M., Felder, G. N., Dietz, E. J. (1988). Learning and teaching styles in engineering education, *Journal of Engineering Education*, 78(7), 674–681. [Full-text article available online at http://www.ncsu.edu/felder-public/Learning_Styles.html; last accessed April 26, 2004]. This article speaks directly to the challenges of teaching and learning chemistry, even though it is written about engineering. The article defines the Felder–Silverman article and describes the *Index of Learning Styles*, now available as a self-scoring web based instrument. Felder discusses the mismatches that exist between common learning styles of engineering students and traditional teaching styles of engineering professors. The article offers solutions and concrete teaching practices that can be implemented to address the full spectrum of student learning styles.

Towns, M. H. (2001). Kolb for chemists: David A. Kolb and experiential learning theory. *Journal of Chemical Education*, 78(8), 1107. This article offers an introduction to Kolb's four learning styles and is written for chemists, by a chemist. Towns describes a learning cycle that can be used to structure any activity so it progresses through each of the four learning styles, including two detailed examples of how the learning cycle can be used to teach both rotational-vibrational spectroscpy and atomic structure and spectra in a physical chemistry course.

References

American Association for the Advancement of Science. (1989). *Project 2061: Science for All Americans* (AAAS Publication 89-01S). New York: Oxford University Press.

Ausubel, D. P. (1963). *The Psychology of Meaningful Verbal Learning*. New York: Grune and Stratton.

Ausubel, D. P. (1968). *Educational Psychology: A Cognitive View*. New York: Holt, Rinehart, and Winston.

Ausubel, D. P., Novak, J. D., and Hanesian, H. (1978). *Educational Psychology*. New York: Holt, Rinehart, and Winston. Reprinted (1986). New York: Werbel and Peck.

Bretz, S. L. (2001). Novak's theory of education: Human constructivism and meaningful learning. *Journal of Chemical Education*. 78(8), 1107.

Donn, J. S. (1990). The relationship between student learning approach and student understanding and use of Gowin's Vee in a college level biology course following computer tutorial instruction. Unpublished doctoral dissertation, Cornell University, Ithaca, NY.

Chapter 3: All Students Are Not Created Equal 39

Edmondson, K. M. (1989). The influence of students' conceptions of scientific knowledge and their orientations to learning on their choices of learning strategy in a college introductory level biology course. Unpublished doctoral dissertation, Cornell University, Ithaca, NY.

Edmondson, K. M., and Novak, J. D. (1993). The interplay of scientific epistemological views, learning strategies, and attitudes of college students. *Journal of Research in Science Teaching*. 30, 547–559.

Entwistle, N. J., and Ramsden, P. (1983). *Understanding Student Learning*. London: Croom Helm.

Felder, R. M. (1993). Reaching the second tier: Learning and teaching styles in college science education. *Journal of College Science Teaching*. 23(5), 286–290.

Felder, R. M., and Henriques, E. R. (1995). Learning and teaching styles in foreign and second language education, *Foreign Language Annals*. 28 (1), 21–31.

Felder, R. M., Felder, G. N., and Dietz, E. J. (2002). The effects of personality type on engineering student performance and attitudes. *Journal of Engineering Education*. 91(1), 3–17.

Felder, R. M., and Soloman, B. A. (2004). Index of Learning Styles. Available online at http://www.ncsu.edu/felder-public/ILSpage.htttml. Last accessed January 2004.

Finster, D. (1989). Developmental instruction. Part I. Perry's model of intellectual development. *Journal of Chemical Education*. 66, 659–661.

Finster, D. (1991). Developmental instruction. Part II. Application of the Perry model to general chemistry. *Journal of Chemical Education*. 68, 752–756.

Flavell, J. H. (1985). *Cognitive Development*. Englewood Cliffs, NJ: Prentice-Hall.

Fleming, N. (2001). *VARK—A Guide to Learning Styles*. Available online at http://www.vark-learn.com/eenglish/index.asp. Last accessed January 2004.

Gardner, H. (1983). *Frames of Mind: The Theory of Multiple Intelligences*. New York: Basic Books.

Gardner, H. (1993). *Multiple Intelligences: The Theory in Practice*. New York: Basic Books.

Gowin, D. B. (1981). *Educating*. Ithaca, NY: Cornell University Press.

Hermann, N. (1990). *The Creative Brain*. Lake Lure, NC: Brain Books.

Jung, C. G. (1971). *Psychological Types*. Princeton, NJ: Princeton University Press.

Kolb, D. A. (1984). *Experiential Learning: Experience as the Source of Learning and Development*. Englewood Cliffs, NJ: Prentice-Hall.

Lawrence, G. (1997). *People Types and Tiger Stripes*. Gainesville, FL: Center for Applications of Psychological Type.

National Research Council. (1996). *National Science Education Standards*. Washington, DC: National Academy Press.

Novak, J. D., and Gowin, B. D. (1984). *Learning How to Learn*. New York: Cambridge University Press.

Perry, W. G. (1968). *Forms of Intellectual and Ethical Development in the College Years: A Scheme*. New York: Holt, Rinehart, and Winston.

Porter, A.; Brophy, J. (1988). Synthesis of research on good teaching: Insights from the work of the institute for research on teaching. *Educational Leadership*. 78–83.

Rickey, D., and Stacy, A. M. (2000). The role of metacognition in learning chemistry. *Journal of Chemical Education*. 77(7), 915–920.

Robertson, M.T. (1984). Use of videotape-stimulated recall interviews to study the thoughts and feelings of students in an introductory biology laboratory course. Unpublished masters' thesis. Cornell University, Ithaca, NY.

Springer, S. P., and Deutsch, G. (1985). Left-brain right-brain (3rd ed). New York: W. H. Freeman.

Tobias, S. (1990). *They're Not Dumb, They're Different: Stalking the Second Tier*. Tucson, AZ: Research Corporation.

Tobias, S. (2002). Learning styles: A challenge for the educator in the classroom. Unpublished paper.

Towns, M. H. (2001). Kolb for chemists: David A. Kolb and experiential learning theory. *Journal of Chemical Education*. 78(8), 1107.

Waterman, M. A. (1982). College biology students' beliefs about scientific knowledge: Foundation for study of epistemological commitments in conceptual change. Unpublished doctoral dissertation, Cornell University, Ithaca, NY.

4

Inquiry and the Learning Cycle Approach

Michael R. Abraham
Department of Chemistry and Biochemistry
The University of Oklahoma

Abstract

Inquiry, as an approach to instruction, is once again making a comeback. It is receiving the attention of teachers at all levels from grade school to college. The National Science Foundation, for example, is encouraging inquiry approaches in the funding of science education initiatives. What inquiry means to different people, however, varies widely. For some, it merely means having hands-on laboratory activities as part of a course. For others, it is a complex learning strategy that pervades all aspects of instruction and curricula. My intention in this chapter is to try and sort out these differences, call upon research and theoretical considerations to identify proven approaches, and provide guidelines for teachers who would like to adopt inquiry into their teaching.

Biography

I am a David Ross Boyd Professor of Chemistry, Adjunct Professor of Science Education, and director of freshman chemistry at the University of Oklahoma. I received a B.A. in Chemistry at Grinnell College, a Masters of Arts in Teaching from Emory University, and a Ph.D. in Science Education from Florida State University. I have taught science at all academic levels from elementary school to college. My research interests are in the field of science/chemical education, specifically instructional strategies, student misconceptions, and the use of computers in helping students visualize atomic and molecular behavior. I have developed curriculum materials using inquiry-oriented instructional strategies at the high school and university levels. At present, I direct the Ph.D. program in Chemical Education at the University of Oklahoma.

Background

In 1902, Alexander Smith, a chemistry professor at the University of Chicago, proposed that the teaching of chemistry should be laboratory based (DeBoer, 1991). His approach was called the Heuristic Method. In this approach, the laboratory played a central role in instruction and was used for two purposes: (1) the verification of chemical principles, and (2) the independent discovery of knowledge. The first use is how laboratory is typically taught in most college chemistry courses today (Abraham et al., 1997). The second represents what most instructors think of when they think of Inquiry as an instructional approach. Smith's approach established the laboratory as a component of instruction in chemistry and the idea that inquiry could be used in chemical education settings.

Inquiry, as an instructional strategy, again became important in the late 1950s with the advent of the National Science Foundation's support of science curriculum reform projects. It was stated that inquiry was a process important to science and as such should be used as a strategy for the instruction of science (DeBoer, 1991, p. 206). It was argued that science instruction should be consistent with the nature of science in order to model scientific processes for students.

Two high school chemistry curriculum projects were originally developed with support from the NSF. They were the Chemical Bond Approach—CBA (Westmeyer, 1961), and the Chemical Education Materials Study—

CHEM Study (Merrill, 1961). The CBA project was more innovative in its instructional approach. The laboratory portion of the course gradually built students toward doing open-ended research-oriented activities by the end of the year. However, the CBA project never caught on. It was felt by many that the approach expected too much from students and the content was too different from the standard chemistry curriculum. As a consequence, although its version of inquiry was a purer example than that of the CHEM Study approach, it didn't have the influence on high school instruction that CHEM Study had. The CHEM Study laboratory program was designed to introduce concepts. Although CHEM Study had a major influence on high school chemistry instruction, the major focus of that influence was not on inquiry-oriented instructional strategies but rather on the revision of the chemistry curriculum toward more modern concepts and principles.

One of the first modern examples of the use of inquiry in college chemistry laboratories was Jay Young's laboratory manual *Practice in Thinking* (Young, 1958). This manual was divided into three parts. Part 1 introduced the student to laboratory procedures and techniques. Part 2 gave the student a group of laboratory actions to perform that illustrated some phenomenon that the student was asked to explain. Part 3 asked the student to design an experiment to demonstrate or prove a chemical idea.

Recently, Inquiry as an instructional strategy, and as an educational outcome of the study of science, has once again become a major emphasis of governmental and professional societies concerned with the improvement of science education at the high school and college levels (Olson and Loucks-Horsley, 2000).

What is Inquiry Teaching?

Inquiry teaching is usually associated with several instructional practices. One of these is the use of the laboratory to introduce concepts rather than verify them. This is sometimes referred to as an inductive use of the laboratory because of its connection to inductive logic—reasoning from specific facts to a generalization. This is in contrast to the more common deductive use of the instructional laboratory—to reason from a known principle to a specific example to verify the principle. A second characteristic of inquiry teaching is the role of the teacher as a guide or facilitator of learning, rather than as source of information. This emphasis manifests itself in the use of questions as the main instructional tactic in the interaction with students. Another characteristic of inquiry teaching is the use of problem-solving activities. Finally, the inquiry teacher will often focus on the processes of science as well as the concepts of science as the goal of instruction.

One of the criticisms of the Inquiry approach is that science has both inductive and deductive functions. As a consequence, it is claimed that the scientific practice of verifying theories (exposition) was being ignored in inquiry teaching (DeBoer, 1991, p. 208; Lawson, 1995, p. 212). Of course, it might be similarly argued that in traditional instruction, the inquiry nature of science was being ignored. Nevertheless, it is reasonable that both characteristics of science should be included in an instructional strategy used to teach science.

Another criticism of Inquiry is its perceived heavy reliance on laboratory as a source of data for generating concepts. It is said that this is an inefficient use of students' time. Joseph Schwab (1963) showed that the Inquiry approach could be based on data sources other than hands-on laboratory. He used verbal presentations of data, called "Invitations to Inquiry," as a starting point for class discussions (see also Gosser, Strozak, and Cracolice, 2001; Moog and Farrell, 2002).

Some have argued that there is no evidence that Inquiry is an effective way to teach. This is simply not true. There is actually an extensive research literature showing that inquiry-oriented teaching strategies have advantages over traditional instructional approaches in attitudes, motivation, and concept and process learning (Lawson, Abraham, and Renner, 1989; Lawson, 1995; Abraham, 1998; Rudd, Greenbowe, and Hand, 2002; Rudd, Greenbowe, Hand, and Legg, 2003). However, as is true of most things in learning and teaching, breaking down the details of this assertion is important for a true understanding of these advantages. (See issue #1 at the end of this chapter.)

Inquiry as Tactic versus Inquiry as Strategy

In the military, a distinction is made between strategy and tactics. Strategy refers to large-scale planning and development to ensure an overall end. Tactics deal with the use and deployment of procedures for obtaining an advantage. The distinction can be seen as a matter of scale. The same distinction can be made between instructional strategies and instructional tactics. Instructors have used the term Inquiry to describe a variety of quite different instructional activities. Using the strategic/tactic distinction, Inquiry can be seen as instructional activities that range all the way from the simple use of questions to the practice of open-ended research. Inquiry as a tactic (e.g., the use of questions) can be seen in the same light as other instructional tactics like the use of concept maps or the group work characteristic of cooperative or collaborative learning. These tactics can be utilized within an overall instructional strategy. An instructional strategy can be seen as the arrangement, combination, and form of learning activities, materials, and instructional tactics designed to meet educational objectives. An inquiry tactic can be used within a lesson or activity while an inquiry strategy would be the overall plan of action for a unit of instruction. Although tactics are important—after all, the success of a strategy depends on using effective tactics—my purpose here is to discuss what I think are some of the issues associated with adopting Inquiry as an instructional strategy.

Instructional Strategies

A useful way of characterizing an instructional strategy used to teach scientific concepts is to divide instruction into phases that play important roles in the instructional process. These include: (1) identification of the concept; (2) demonstration of the concept; and (3) application of the concept. Many science educators have recommended instructional strategies consisting of these phases, although some have subdivided the phases into more components or have added additional components like evaluation (Karplus and Thier, 1967; Torrance, 1979; Hewson, 1981; Renner, 1982; Bybee and Landes, 1990; Rudd, Greenbowe, Hand, and Legg, 2001). Differences among instructional strategies can be characterized by three characteristics: whether all of the phases are included in the unit of study, how the three phases are arranged (i.e., their sequence), and the formats of the activities in each of the phases: laboratory, discussion, lecture, readings, problem sets, etc. (Abraham 1988–89).

In order to illustrate these ideas, two instructional strategies will be compared with regard to phases of instruction (Renner 1982).

Traditional (Concept → Data)

Phases of Instruction	Goal	Activities	Questions	Data
Inform	Present Concept	Lecture / Discussion, Readings	What is the concept?	
Verify Concept	Confirm the truth of concept	Laboratory, Demos	How do your observations fit the concept?	Confirm Concept with data, Provide Evidence
Practice Concept	Apply, reinforce, review, extend, and understand concepts	Readings, Problem Sets, Application Questions,	Using what you know, answer the following…	
Evaluate		Examinations, Quizzes		

Figure 1. Characteristics of Traditional Instruction.

The Traditional Approach to General Chemistry

At the present time, general chemistry is taught in colleges and universities in a fairly uniform way. As represented in Figure 1, this teacher-centered instructional strategy can be seen as being divided into phases that are taken in order. First, students are assigned readings in a textbook, are expected to attend lecture where often the same material is presented, and listen passively while taking notes. This might be called the "Inform Phase" and is used to identify the concept. This phase is followed by a laboratory activity that either verifies the concept that students were already informed about in lecture (called the "Verification Phase," which demonstrates the concept), or is completely unrelated to what is being covered in lecture. Then, students are assigned problems from the end of the chapter in the textbook (called the "Practice Phase," which applies the concept). After a period of time, they are given an examination to test what they have learned. The "Inform—Verify—Practice" (I—V—P) sequence corresponds to the three phases previously discussed. For many reasons, not the least of which is the negative attitude of students, many chemistry instructors have become dissatisfied with this approach.

The Learning Cycle Approach

An alternative instructional strategy is a student-centered inquiry-oriented approach called the Learning Cycle (Lawson, Abraham, and Renner, 1989; Lawson, 1995; Marek and Cavallo, 1997). The Learning Cycle approach, as represented in Figure 2, can also be seen as being divided into phases that are taken in order. First, students are exposed to data (called the "Exploration Phase," which demonstrates the concept) from which concepts can be derived (called the "Invention Phase," which identifies the concept). Students can then apply the concept to other phenomena (the "Application Phase," which applies the concept). In contrast to the traditional approach, this inquiry-oriented approach is based upon data. This difference has several consequences for the role played by various instructional activities. Laboratory and other data generating activities play a more central role in instruction by introducing concepts rather than verifying concepts. The curriculum can be said to be data driven. Classroom discussions are focused on using data to generate concepts rather than informing students of the concepts. Textual materials are used to apply, reinforce, review, and extend concepts rather than introduce concepts. This approach encourages more active learning by students.

Inquiry (Data → Concept)

Phases of Instruction	Goal	Activities	Questions	Data
Explore	Explore relations and patterns in data	Laboratory, Demos, MoLES, Lab Simulations, Video	What did you do? What did you observe?	Gathering Data
Invent Concept	Develop and understand concepts with teacher/peers	Lecture / Discussion	What does it mean?	Explaining Data
Apply Concept	Apply, reinforce, review, extend, and understand concepts	Readings, Problem Sets, Application Questions, Verification Laboratory	Using what you know, answer the following…	Using Data, Provide Evidence
Evaluate		Examinations, Quizzes		

Figure 2. Characteristics of the Learning Cycle Approach.

The Learning Cycle approach is an inquiry-based instructional strategy derived from constructivist ideas of the nature of science (Bodner, 1986), and the developmental theory of Jean Piaget (Piaget, 1970). Although Piaget's theories are too complex to discuss in detail here, a brief consideration of one aspect of his ideas is provided to clarify how the Learning Cycle approach is consistent with these ideas. According to Piaget, human beings have mental structures that interact with the environment. We assimilate or transform information from our environment into our existing mental structures. Our mental structures operate on the assimilated information and transform it in a process of accommodating to it. Thus, information from the environment transforms our mental structures, while at the same time, our mental structures transform the information. This change is driven and controlled by the process of disequilibration. If there is an incompatibility between the assimilated information and existing mental structures, disequilibration takes place. This requires a change or accommodation of the mental structure or a change in the perception of the assimilated information. When our mental structures have accommodated to the assimilated information, we are in a state of equilibrium and have reached an "accord" of thought with things" (Piaget, 1963, p. 8). In accommodating the new information, however, the altered mental structure can become disequilibrated with related existing mental structures. The new structure must be organized with respect to the old structures to develop a new equilibrated organization. In other words, we must bring the "accord of thought with itself" (Piaget, 1963, p. 8). This overall process, called Piaget's functioning model, has implications for instruction. (For an overview of Piaget's theories, refer to Dudley Herron's *Journal of Chemical Education* article, "Piaget for Chemists" (Herron, 1975), and Good, Mellon, and Kromhout's article "The Work of Jean Piaget" in the same journal (Good, Mellon, and Kromhout, 1978).)

If learning spontaneously occurs through a process of assimilation accommodation and organization, then instruction could take advantage by sequencing instructional activities to be compatible with the nature of learning. In order to facilitate assimilation, instructional activities should expose the learner to a segment of the environment that demonstrates the information to be accommodated. This should be followed by activities that help the learner to accommodate to the information. Finally, in order to organize the accommodated information, activities should be developed to help the learner to see the relation between the new information and other previously learned information. The parallels between Piaget's functioning model, the Learning Cycle approach, and learning activities are illustrated in Figure 3.

Piaget's Functioning Model	Learning Cycle Teaching Model		Learning Activities and Materials
Assimilation	Exploration	*E*	Data Collection and Analysis
Accommodation	Concept Invention	*I*	Conclusions and/or Interpretation
Organization	Application	*A*	Application Activities

Figure 3. Piaget Functioning Model and the Learning Cycle Approach.

There are several characteristics which, when used in combination, establish the Learning Cycle approach as a distinct instructional strategy. The most important of these is the presence of the three phases of instruction in a specific sequence: "exploration—concept invention—concept application" (E—I—A). This sequence has a number of logical consequences. The "Exploration Phase" coming first implies that learners will use the information gained during the learning activity to propose or invent an explanation. This is an inductive use of data (proceeding from the specific to the general). The key to this instructional approach is that the learner derives the concept from their observations of the behavior of a chemical system. In this sense, data plays a central role in instruction. In the "Application Phase," learners use the invented concept to verify and modify their ideas. This is a deductive use of data (proceeding from the general to the specific). The learning cycle approach has advantages over other instructional strategies because it takes into account both inquiry and exposition. That is, it requires the learner to use both inductive and deductive logical processes.

There has been a large amount of research concerning the Learning Cycle approach since its origins in the 1960s. Most of the research supporting the Learning Cycle approach is discussed in detail in Lawson, Abraham, and Renner (1989). A summary of this research supports the conclusion that the Learning Cycle approach can result in greater achievement in science, better retention of concepts, improved attitudes toward science and science learning, improved reasoning ability, and superior process skills than would be the case with traditional instructional approaches (see Raghubir, 1979; Renner, Abraham, and Birnie, 1985; Abraham and Renner, 1986; Ivins, 1986; McComas III, 1992). This is especially true with intermediate level students where instructional activities have a high level of intellectual demand (Lott, 1983).

Issues in the use of Inquiry as an instructional strategy

Issue #1— What are you trying to teach? "There is no one best way to teach" is an educational cliché that, although having elements of truth, is so general that it offers no guidance to practicing teachers. This cliché might be incorrectly interpreted to mean that it makes no difference how one teaches. A better interpretation is that different learning outcomes are best approached by different instructional means (Gagné and Briggs, 1979; Gagné, 1985). A useful way to address this dilemma is to divide the learning outcomes of a chemistry course into categories. Figure 4 lists these outcomes as concepts, processes, skills, facts, and attitudes. It is useful to do this because different specific instructional tactics have been shown by research to be educationally effective for different categories of learning.

Category of Learning	Definition	Example	Instructional Tactic
Concepts	Generalization, Principle, or Theory	Conservation of Mass	Inquiry—Questioning
Processes	Method	Separation and Control of Variables	Practice
Skills—Laboratory —Mathematical	Ability	Using a Balance Curve Fitting	Informing or Demonstrating
Facts	—Observation —Definition	$Cu^{2+}(aq)$ is blue Ag is silver	Observing Informed in Context
Attitudes	Beliefs or Feelings	Chemistry is Fun	Example

Showing

Figure 4. Effective Instructional Tactics.

Concepts are theories or principles that are used to explain phenomena. Research has shown that inquiry tactics (e.g., the use of questioning) are the best instructional tactics for concept learning. In traditional instructional strategies, concepts are often treated as facts. As a consequence, less effective instructional tactics (e.g., lectures) are commonly used.

Processes are methods. The science education literature has developed a long list of these processes. A representative list can be found in Table 1 (Livermore, 1964). Research has shown that scientific processes are best learned by practicing them in multiple settings.

There are two kinds of skills of interest: laboratory skills and mathematical skills. Unlike processes, skills are more specific to particular situations. However, there is some overlap between processes and skills. Skills are best learned by being shown how to use the skill by an expert.

Facts are truths. They are true either by definition or by observation. Although teachers often require their students to memorize facts, a more effective approach is to teach facts in context by using them to develop concepts. Students find it easier to remember factual information if related facts are associated with a concept or principle. *– Like words in a song.*

Finally, attitudes are learned through example *(by instructor)*.

The Learning Cycle approach gives an instructor the opportunity to expose students to all of these categories of learning. Learning cycles are generally built around concepts. These concepts are usually introduced in a laboratory setting or some other source of data from which the concept can be developed through discussion. The collection, manipulation, and interpretation of data give students ample opportunity to practice scientific processes. In the laboratory, students learn how to use equipment and procedures. In processing data, they learn various mathematical skills. During their observations and discussions, they are exposed to factual information connected with the topic they are studying. In using the ideas that they gain from their activities, they can again use processes to expand their understandings.

Table 1—Scientific Processes

Basic Processes	Integrated Processes
Observing	Hypothesizing
Using Time/Space Relationships	Controlling Variables
Orienting	Interpreting Data
Measuring	Reasoning (Inductive and Deductive)
Classifying	Drawing Conclusions
Identifying and Differentiating	Evaluating
Discriminating	Defining Operationally
Communicating	Experimenting
Describing	Modeling
Comparing	
Predicting	
Inferring	

Issue #2—How do you design inquiry-oriented activities? Following is a set of guidelines for constructing inquiry activities using the Learning Cycle approach.

1) Identify the concept/principle/law that you are trying to teach as the target of the activity. This is the tricky part. This should be a big idea, not a skill or fact. The activity should concentrate on a single concept, certainly not more than two. If more than two concepts are involved, you should subsume them under a larger and more central concept, put them in separate activities, or address them in follow-up application activities related to the main concept.

2) Write a concept statement. This should be a brief description of the concept to be taught.

3) Write a problem statement/question. This should be a descriptive statement or question whose answer leads to the concept. Be careful to not give the concept away in the statement. These statements can be used to introduce the activity to students.

4) Identify the data/observations that can be used to explore the concept. Write procedures that will cause students to collect that data and/or make the observations.

5) If necessary, write procedures that will cause students to organize the data into a form that will facilitate interpretation (e.g., in tabular or graphical form).

6) Write questions or procedures that will lead the student to interpret the data or to draw a conclusion that will develop the target concept.

7) Write questions or activities that will lead the student to use or apply the concept in a new setting.

8) (Optional) Ask students to represent, model, or visualize the invented concept at the molecular level by drawing a diagram(s) that explains/represents/models their observations.

9) (Optional) Write questions that ask students to design experiments that will answer these questions. These can be used to cause students to explore related concepts, reinforce the concept, or apply the concept. Use anticipated or observed student misconceptions to generate open-ended challenges (e.g., "prove or disprove the following statement...").

Pre-laboratory activities should be limited to safety in the laboratory and procedures and skills necessary to do the activity. It should not tell the students what the concept to be learned is, how to do the laboratory, what they will observe, or what they will conclude.

As an example of a simple learning cycle lesson, consider the following instructional activity designed to teach the concept of acids and bases (Abraham and Pavelich, 1999a, p. 99). The concept statement might be "Acids are any species that produces H^+ ions in water solution. Bases are species that produce OH^- ions in water solution." The problem statement might be "What are the characteristics of acid and base solutions?" Students are provided with a number of unknown solutions and asked to perform tests that might include the use of several acid/base indicators, the interaction of the solutions with active metals (which react with acids), and interaction of the solutions with soluble metal ion solutions (which precipitate in bases). They might also be asked to measure the conductivity of each solution. During this "Exploration Phase" of instruction, students are asked to identify patterns that enable the unknown solutions to be grouped. During a subsequent class discussion, the students would compare data and notice that the solutions could be classified into three groups. The instructor then might define these groups operationally and label them as acids, bases, and neutral substances ("Concept Invention Phase"). Depending on the age and scientific sophistication of the students, the instructor might continue the discussion by introducing the chemical formulas of the solutions and ask what the formulas in each category have in common. This information could be used to invent a theoretical definition of acids and bases. Using these ideas, students might then be given "Application Phase" activities involving additional solutions to fit into the classification scheme. They also might be asked to rationalize the acid/base characteristics of substances that don't seem to be consistent with the previously invented acid/base theory.

Issue #3—What kind of questions should be used? One of the most important tactics of inquiry instruction is the use of questions (Dillon, 1983; Saunders and Shepardson, 1987; Bateman, 1990; Mazur, 1997). A way to analyze this tactic is to classify questions according to type in order to guide their use. A useful classification scheme is discussed by Pavelich (1982) and is based on the level of thinking processes necessary for students to respond.

Another way of addressing questioning is to look at what kinds of questions can be used with data driven activities like those of a learning cycle (see Table 2). Questions used by an instructor during or right after an exploration activity might be *What did you do?* and *What did you observe?* These questions serve a useful role in a class discussion. The *What did you do?* question serves to orient the students to the focus of the discussion. The *What did you observe?* question establishes consensus concerning what happened, allows students to resolve differences, and encourages a focus on the data that will be used to invent the target concept. During the "Concept Invention Phase," the question type used by the instructor is *What does it mean?* This question allows students to invent a concept or have a concept invented for them by the instructor. Finally, during the "Application Phase," students can be asked to answer questions that require the concept as prerequisite knowledge.

Issue #4—How do instructional tactics fit into the Learning Cycle Strategy? Other instructional tactics besides questioning can be utilized in a learning cycle. Student generated concept maps are a research-proven method for helping students to organize their knowledge of a concept with other related concepts (Novak and Gowin, 1984; Novak and Wandersee, 1990). As such, the construction of a concept map is an excellent "Application Phase" activity. (See Chapter 11.)

Cooperative learning and other small group tactics are also an excellent research-proven replacement for more formal teacher-led discussions and lectures (Johnson, 1991; Johnson, Johnson, and Holubec, 1993). Cooperative procedures can also be used in laboratory (Cooper, 2003: Abraham and Pavelich, 1999b, p. 13). Cooperative learning activities can be used in many of the phases of the learning cycle approach. As "Exploration Phase" activities, they can be used by students to explore the *What did you observe?* question (for a non-laboratory example, see Moog and Farrell, 2002); as "Concept Invention Phase" activities to organize their data and begin to address the *What does it mean?* question, and as "Application Phase" activities to address extensions of their concept (e.g., Gosser, Strozak, and Cracolice, 2001).

The science writing heuristic (Rudd, Greenbowe, and Hand, 2002; Rudd, Greenbowe, Hand, and Legg, 2003) can also be used as a format for the laboratory portion of a course. This approach encourages students to formalize their inquiry into chemical systems by giving students an inquiry-based procedure for planning and reporting their chemical investigations.

Issue #5—Guided versus Open Inquiry. Inquiry-oriented instructional strategies are usually placed in one of two categories, guided or open. A useful way to look at this distinction was first proposed by Pella (1961) and was used to distinguish types of laboratory activities. Figure 5 is based on Pella's vision. This figure identifies components of instructional laboratory work and asks the question, *Who is in charge of the decision making for that component, the teacher (T) or the student (S)?* In the traditional or verification laboratory, the teacher is in charge of making all of the decisions. In open inquiry, the student is in charge of all of the decisions. In guided inquiry, the teacher and student share the responsibility for decision-making.

	Verification	Guided Inquiry	Open Inquiry
Choose Problem	T	T	S
Design Experiment	T	T	S
Collect Data	T	S	S
Interpret Results	T	S	S

Figure 5. Degree of Freedom in Instructional Strategies.

In a research study (Abraham, 1982), students were exposed to the three laboratory types in Figure 5 and asked to describe the main characteristics of their laboratory experience by choosing from a list of 25 descriptive statements. According to these students, the major goal of the verification laboratory was to develop skills in the techniques and procedures of chemistry. Students said they followed step-by-step instructions, recorded information required by the instructions, and discussed their data and conclusions with each other. Laboratory was characterized as requiring interpretation of data and answering specific questions. Instructors were said to be mostly concerned with the correctness of the data collected by students. Students in guided inquiry laboratory said they followed step-by-step instructions, answered specific questions, and discussed and explained laboratory phenomena. These students said that laboratory reports required them to provide evidence to back up their conclusions. Students said that open inquiry laboratory required them to back up their conclusions with evidence. Students also said they designed their own experiments.

Using these ideas, the distinction between guided and open inquiry can be characterized as the degrees of freedom in decision-making allowed to students. Open inquiry activities can play a useful part in a learning cycle. For example, an "Application Phase" laboratory activity could ask the student to design and carry out an experiment to investigate an application of the concept they developed earlier (Abraham and Pavelich, 1999a, p. 275). If a student developed a misconception associated with an investigation, he/she might be challenged to design an experiment to prove or disprove their conception.

Summary

The Learning Cycle Approach is an inquiry-oriented instructional strategy that has great promise for chemistry instruction. It has a solid research base testifying to its effectiveness. It has been shown to be superior to traditional instructional approaches developing achievement in science, retention of concepts, improved attitudes toward science and science learning, improved reasoning ability, and process skills. This has been shown to be especially true with intermediate level students where instructional activities have a high level of intellectual demand. The Learning Cycle Approach gives guidance to instructors as to how to interact with students during instruction, how to design activities for classroom use, and what to emphasize as the goal of instruction. A wide variety of proven instructional tactics can be utilized within its format.

Suggestions for Further Reading

To learn more about Inquiry and the Learning Cycle Approach, the author recommends the following references:

Abraham, M. R. (1988–89). Research on instruction strategies. *Journal of College Science Teaching*. 18(3), 185–187, 200.

Allen, J. B., Barker, L. N., et al. (1986). Guided inquiry laboratory. *Journal of Chemical Education*. 63(6): 533–534.

Pavelich, M. J. and Abraham, M. R. (1979). An inquiry format laboratory program for general chemistry. *Journal of Chemical Education*. 56(2): 100–103.

References

Abraham, M. R. (1982). A descriptive instrument for use in investigating science laboratories. *Journal of Research in Science Teaching*.19(2), 155–165.

Abraham, M. R. (1988–89). Research on instruction strategies. *Journal of College Science Teaching*. 18(3), 185–187, 200.

Abraham, M. R. (1998). The learning cycle approach as a strategy for instruction in science. In K. Tobin and B. Fraser (Eds.), *International Handbook of Science Education*.513–524. The Netherlands, Kluwer.

Abraham, M. R., Cracolice, M. S., Graves, A. P., Aldahmash, A. H., Kihega, J. G., Palma Gil, J. G., and Varghese, V. (1997). The nature and state of general chemistry laboratory courses offered by colleges and universities in the United States. *Journal of Chemical Education*. 74(5): 591–594. (Also available online at http://jchemed.chem.wisc.edu/).

Abraham, M. R. and Pavelich, M. J. (1999a). *Inquiries into chemistry, 3rd ed.* Prospect Heights, IL: Waveland Press.

Abraham, M. R. and Pavelich, M. J. (1999b). *Inquiries into chemistry: Teacher's guide, 3rd ed.* Prospect Heights, IL: Waveland Press.

Abraham, M. R. and Renner, J. W. (1986). The sequence of learning cycle activities in high school chemistry. *Journal of Research in Science Teaching*. 23(2), 121–143.

Allen, J. B., L. N. Barker, et al. (1986). Guided inquiry laboratory. *Journal of Chemical Education*. 63(6): 533–534.

Bateman, W. L. (1990). *Open to question: The art of teaching and learning by inquiry*. San Francisco, Jossey-Bass.

Bodner, G. M. (1986). Constructivism: A theory of knowledge. *Journal of Chemical Education*. 63(10), 873–878.

Bybee, R. W. and Landes, N. M. (1990). Science for life and living. *The American Biology Teacher*. 52(2), 92–98.

Cooper, M. M. (2003). *Cooperative chemistry: Laboratory manual, 2nd ed*. Boston: McGraw-Hill.

DeBoer, G. E. (1991). *A history of ideas in science education*. New York: Teachers College Press.

Dillon, J. T. (1983). *Teaching and the art of questioning*. Bloomington, Indiana: Phi Delta Kappa Educational Foundation.

Gagné, R. M. (1985). *The conditions of learning and theory of instruction*. New York: Holt.

Gagné, R. M. and Briggs, L. J. (1979). *Principles of instructional design*. New York: Holt.

Good, R., Mellon, E. K., and Kromhout, R. A. (1978). The work of Jean Piaget. *Journal of Chemical Education*. 55(11): 688–693.

Gosser, D. K., Strozak, V. S., and Cracolice, M. S. (2001). *Peer-led team learning: General chemistry*. Upper Saddle River, NJ: Prentice Hall.

Herron, J. D. (1975). Piaget for chemists. *Journal of Chemical Education*. 52(3): 146–150.

Hewson, P. W. (1981). A conceptual change approach to learning science. *European Journal of Science Education.* 3(4), 383–396.

Ivins, J. E. (1986). A comparison of the effects of two instructional sequences involving science laboratory activities, University of Cincinnati. 46(8), 2254A. Dissertation abstracts.

Johnson, D. W. (1991). *Active learning: Cooperation in the college classroom.* Minneapolis, MN: Burgess.

Johnson, D. W., Johnson, R. T., and Holubec, E. J. (1993). *Cooperation in the classroom.* Minneapolis, MN: Burgess.

Karplus, R. and Thier, H. D. (1967). *A new look at elementary school science.* Chicago: Rand McNally.

Lawson, A. E. (1995). *Science teaching and the development of thinking.* Belmont, CA: Wadsworth Publishing Company.

Lawson, A. E., Abraham, M. R., and Renner, J. W. (1989). *A theory of instruction: Using the learning cycle to teach science concepts and thinking skills [Monograph, Number One].* Kansas State University, Manhattan, KS: National Association for Research in Science Teaching.

Livermore, A. H. (1964). The process approach of the AAAS commission on science education. *Journal of Research in Science Teaching.* 2(4), 271–282.

Lott, G. W. (1983). The effect of inquiry teaching and advanced organizers upon student outcomes in science education. *Journal of Research in Science Teaching.* 20(5), 437–451.

Marek, E. A. and Cavallo, A. M. L. (1997). *The learning cycle: Elementary school science and beyond.* Portsmouth, NH: Heinemann.

Mazur, E. (1997). *Peer instruction: A user's manual.* Upper Saddle River, New Jersey: Prentice Hall.

McComas III, W. F. (1992). The nature of exemplary practice in secondary school science laboratory instruction: A case study approach, University of Iowa. 52(12), Dissertation abstracts. 4284A.

Merrill, R. J. (April 1961). Chemistry: An experimental science. *The Science Teacher.* 26–31.

Moog, R. S., and Farrell, J. J. (2002). *Chemistry: A guided inquiry, 2nd ed.* New York, Wiley.

Novak, J., and Wandersee, J. E. (Eds.) (1990). Concept mapping (Special Issue). *Journal of Research in Science Teaching.* 27(10).

Novak, J. D. and Gowin, D. B. (1984). *Learning how to learn.* Cambridge, England: Cambridge University Press.

Olson, S. and Loucks–Horsley, S. (Eds.) (2000). *Inquiry and the national science education standards: A guide for teaching and learning.* Washington, DC: National Academies Press.

Pavelich, M. J. (1982). Using general chemistry to promote the higher level thinking abilities. *Journal of Chemical Education:* 59(9), 721–724.

Pavelich, M. J. and M. R. Abraham (1979). An inquiry format laboratory program for general chemistry. *Journal of Chemical Education:* 56(2): 100–103.

Pella, M. (September 28, 1961). The laboratory and science teaching. *The Science Teacher.* 20–31.

Piaget, J. (1963). *The origins of intelligence in children.* New York: Norton.

Piaget, J. (1970). *Structuralism*. New York: Harper and Row.

Raghubir, K. P. (1979). The laboratory-investigative approach to science instruction. *Journal of Research in Science Teaching*. 16(1), 13–18.

Renner, J. W. (1982). The power of purpose. *Science Education*. 66(5), 709–716.

Renner, J. W., Abraham, M. R., and Birnie, H. (1985). The importance of the form of student acquisition of data in physics learning cycles. *Journal of Research in Science Teaching*. 22(4), 303–325.

Rudd, II, J. A., Greenbowe, T. J., and Hand B. M. (2002). Restructuring the chemistry laboratory notebook using the science writing heuristic. *Journal of College Science Teaching*. 31(4), 230–234.

Rudd, I., J. A., Greenbowe, T. J., Hand, B. M., and Legg, M. J. (2001). Using the science writing heuristic to move toward an inquiry-based laboratory curriculum: An example from physical equilibrium. *Journal of Chemical Education*. 78(12), 1680–1686.

Saunders, W. L., and Shepardson, D. (1987). A comparison of concrete and formal science instruction upon science achievement and reasoning ability of sixth-grade students. *Journal of Research in Science Teaching*. 24(1), 39–51.

Schwab, J. J. (1963). *Biology Teachers Handbook*. New York: Wiley.

Torrance, E. P. (1979). A three-stage model for teaching creative thinking. *The psychology of teaching for thinking and creativity: 1980 AETS yearbook*. In A. E. Lawson (Ed.) Columbus, Ohio, ERIC/SMEAC and the Association for the Education of Teachers in Science. 226–253.

Westmeyer, P. (May 1961). The chemical bond approach to introductory chemistry. *School Science and Mathematics*. 317–322.

Young, J. A. (1958). *Practice in thinking: A laboratory course in introductory chemistry*. Englewood Cliffs, N. J.: Prentice-Hall.

Relevance and Learning The...

Donald J. Wink
Department of Chemistry
University of Illinois at Chicago

Abstract

The concept of relevance is used widely in teaching and in the development of instructional materials. Relevance is a goal of many projects, even emerging as an organizing principle in some cases. However, there are theoretical problems inherent in the idea of making something relevant, and these problems can appear as barriers to student engagement. Learning theory includes important ideas about student-centeredness, meaning-making by students, and student openness to learning that frame a set of practical elements that are important to "relevant" teaching: the particular student, the mediator, the materials, and assessment. Examples of how these practical elements are addressed in the science education literature, with particular emphasis on chemistry, are presented. Models of teaching that incorporate multiple elements are also presented.

Biography

Donald J. Wink is Professor and Head in the Department of Chemistry at the University of Illinois at Chicago. He has engaged in several materials and curriculum development projects since he arrived there for a faculty position as Coordinator of General Chemistry in 1992. Prior to that, he was an assistant professor at New York University engaged in research in theoretical, synthetic, and applied organometallic chemistry. He also co-developed a project-based physical chemistry lab program organized around the synthesis and characterization of porphyrin complexes. He was trained for this program at the University of Chicago (S.B.) and at Harvard University (Ph.D.) His current projects are diverse but share a theme of crossing boundaries, often using student pathways as a source of inspiration and direction. His first UIC project joined preparatory chemistry and intermediate algebra curricula in a curriculum development and research project that demonstrated gains for student outcomes in later chemistry classes (Wink et al., 2001) and the publication of a new "math-aware" preparatory chemistry text, *The Practice of Chemistry*. A later project, discussed in this paper, involved faculty from other departments that require general chemistry in the development of scenario-based laboratory instruction. His most recent work focuses on issues of teaching in K–12 settings, including a collaborative effort for teacher preparation that brought together UIC and area community colleges and an NSF GK–12 project for intervention in schools. As part of the latter project, he is a regular participant in activities at Crane Tech Prep High School on Chicago's West Side, where he works with a community of administrators, teachers, and students addressing some of the most challenging teaching and learning issues of urban schools. Not surprisingly, he relies on a network of valuable colleagues and coauthors within the greater Chicago area.

Introduction

We have all heard the terms, "real-world," "relevant," and "student-centered." These words evoke classrooms that do more than just present the material. They offer the promise of learning experiences that give special attention to the past, present, or even future viewpoint of students. These experiences revolve around specific events to which students, in theory, attach positive personal meaning, enhancing their engagement in the lesson and their learning of content.

...as with so many ideas, is easy to recognize because it causes dramatic changes in how a student ... One evocative example of this, though not derived from science education, is in Frank McCourt's ... autobiography *'Tis*. McCourt, better known as the author of *Angela's Ashes*, came to America and ...ed as a teacher. At a vocational high school where neither students nor teachers had little engagement, he ...appened upon a decade-old trove of ungraded papers in his classroom closet:

> *I pile the crumbling pages on my desk and begin reading to my classes. They sit up. There are familiar names. Hey, that was my father. He was wounded in Africa. Hey, that was my Uncle Sal that was killed in Guam.*
> *When I read the essays aloud there are tears. Boys run from the room to the toilets and return red–eyed. Girls weep openly and console one another.*
> *They are suddenly interested in compositions with the title "My life..."*

Relevance and Chemical Education

McCourt was lucky; educators more commonly set out to add relevance in a planned way. This is found again and again in descriptions of curriculum innovation. The National Science Foundation's Course, Curriculum, and Laboratory Improvement program has certainly been a major source of funding for relevant ideas: a recent database search covering the last five years found 46 grants worth $4.7 million that contain "chemistry" and either "relevance or relevant" in their abstract. Most textbooks in introductory chemistry are rife with little side-bars that bring in ideas labeled as relevant. In fact, the American Chemical Society has organized three texts around relevance: *Chemistry in the Community* for high school (ACS, 2001), *Chemistry in Context* (ACS, 2003) for general education college students, and the general chemistry textbook *Chemistry* (ACS, 2004).

The author and his coworkers were also among those who combined materials development with relevance in an NSF proposal under the title, "The Chemical Professional Laboratory Program" or CPLP. This has now been commercialized as "*Working with Chemistry*" (WWC; Wink et al., 2005). This laboratory curriculum includes a scenario for each experiment group drawn from actual uses of chemistry concepts in *other* professions, ranging from nursing to mechanical engineering. As we said in the body of that proposal:

> *Our primary goal is to make the laboratories an environment where students learn chemical techniques along with the relevance of chemical procedures to the workplace. This should help students to pay more attention to the relevance of their training. We feel it is not enough to say 'this lab involves stoichiometry.' Instead, through this program, we hope to show students both the study of stoichiometry and its context in the analysis of phosphate in a soil sample gathered by an field ecologist.*

We had practical reasons to develop a "relevant" curriculum. We were working at a broad research university and a closely linked community college, so we were able to collaborate with faculty on our own campus who taught in those professional programs that require chemistry. And our students were intent on careers scattered across the university's colleges. For example, a physiology researcher in nursing helped us put together a program to teach acid-base buffers. The relevant context for this was the restoration of a blood's pH in the wake of a heart attack. The nursing professor gave us a deep understanding of physiological pH. And we expected that learning how to "save a patient" would seem very relevant to all students, especially those planning on a health career.

Our goals for this curriculum were also conceptual. For example, we cited the late Orville Chapman, a leader in curriculum reform over the last 15 years (Russell et al., 1998). At an important moment in his projects, he wrote (Chapman, 1993), "We have no hope of expanding our clientele with our present structure. The structure is not sound. The message from 80% of college students comes in loud and clear. Chemistry without people, economics, and policy is irrelevant. We choose to be irrelevant; they ignore us." We expected that, as with many other relevant curricula, we would see increased student attention.

We set out to make chemistry laboratory work more relevant to students by selecting scenarios that might let them act out events from a future career based on chemistry. We found that it was easy and informative to do this; along with our network of co-developers and testers, we learned a lot about how chemistry is relevant in other programs, such as medical technology, nursing, soil chemistry, etc. For example, we were also able to learn the relevance of acid-base titration to field ecology. A biology collaborator took us through how he studied carbon element cycles through the selective precipitation of barium carbonate prior to an acid-base titration of solution hydroxide. This kind of work made it fun for the authors and their collaborators to create these relevant materials. But we found that students do not always find the scenarios engaging in a way that led them to devote the extra time to working with the scenario instead of just "doing" the chemistry. For example, a lab on decomposition kinetics required them to try different methods before settling on one, just as a chemical engineering researcher would. Instead, they would look to just set up an apparatus, take measurements, and try later to do what they could to make it fit an ideal model. We are also aware that the program works in very different ways at a large, public, urban commuter campus and at a suburban community college.

The reasons for the barriers to student engagement are complex, but they are covered concisely in a discussion presented by Karen Barad (2000), a physicist and a professor of philosophy and women's studies.

> *...there's something paradoxical about the notion that something can be "made" relevant—as if relevancy could be imposed or added onto an existing structure. The starting point is all wrong: taking an existing course and contemplating superficial alterations to it in an effort to make it relevant is a poor substitute for designing a relevant course or curriculum. Another important limitation is that "relevancy" is undertheorized. "Relevant for/to whom?" is a crucial question that needs to be addressed; but it is insufficient if posed in isolation from the issue of how the "whom" is to be understood. Questions of relevancy are intertwined with questions of subjectivity and epistemic responsibility. For example, although the physics teacher who creatively suggests a "kitchen physics" approach for the "ladies" has clearly thought about the fact that relevancy has different meanings for different audiences, his approach is essentializing: it fixes what is considered "feminine" in particular ways".*

Barad's paper prompted us to ask how well we had considered the question of "relevant to/for whom?" We also knew that we, too, might be "essentializing" students. That is, we might be putting them into a context we *think* they will like because of our assumptions about their "essence" as learners. If this were the case, then we can't be surprised if they reject our assumptions and the standpoint of our writing entirely.

Success and Challenges in Relevance in Science and Chemistry Curricula

The literature contains several reports which demonstrate that student engagement with relevant instruction can be quite high and consistent in some cases—for example, when popular films are used for teaching. James Goll and his coworkers have studied this with the film *Apollo 13* (Goll and Woods, 1999) and the series *From the Earth to the Moon* (Goll and Mundinger, 2003). They reported how the novelty, Hollywood-quality graphics, and evocative scenes quickly engage students in discussions of what science and chemistry is relevant in a particular situation. For example, *Apollo 13* contains a dramatic depiction, with appropriate chemical terms, of the crisis of carbon dioxide buildup when the spacecraft's scrubbers were overwhelmed. This has also been studied from a more systematic research perspective in middle school science (Sherwood et al., 1987). In that case, clips from *Indiana Jones and the Raiders of the Lost Ark* were used to create a "macro-context" for learning (Bransford et al., 1990; Cognition and Technology Group at Vanderbilt, 1993). The clip they use includes Jones' foolish attempt to replace a solid gold statue with a bag of sand, showing the importance of density. This study found improved student learning of information about biology and density after even a one-hour intervention in middle school. The point was made that these films become a context for mediation, where the instructor plays an essential part in ensuring that students actually see the film as a place to learn science, not just as a pleasant diversion. More recent reports include work on the use of shorter clips in general chemistry (Wink 2001a), such as the chlorine gas attack in the film *Legends of the Fall*.

An important point about these reports, especially the careful research in the *Raiders* paper, has to do with linking particular clips to learning through the intervention of the instructor. As Sherwood and coworkers put it, "A major purpose of the videos is to provide a context for mediation. A mediator's role is not simply to act as a stimulus,...but instead to provide structure for the experiences of the learner." They pointed out that, in the absence of a mediator, the clips entertain students but do not support learning *per se*.

There have also been extensive studies of the impact of the relevance-centered instruction of *Chemistry in the Community* on student learning. This high school text notably presents less material than many other books, omitting, for example, molecular orbitals and equilibrium calculations. However, the book may be more effective over time, as suggested in a study comparing a *ChemCom*-based course to a course based on a traditional text (Winthur and Volk, 1994).

Teaching with relevance is not, however, automatically successful. As mentioned, the CPLP/WWC materials are a place where relevance sometimes did not always work well. Observations pointed to the role of the instructor over time as critical to students' engagement or outright rejection of the relevant scenario. When the materials were taught in a community college, the scenarios became a common discussion point for the class. The instructor carried out the kind of mediation Sherwood et al. mention, where the scenario is cited frequently before, during, and after the lab itself. On the other hand, faculty working with teaching assistants at a large university encountered a different situation. The assistant and the students did actively engage in discussion over time, but these focused on how to design the experiment efficiently, with little of the inquiry or the relevance on which the program is based.

The mediation problem has also been mentioned in connection with one of the major systemic efforts to make curricula relevant: the *ChemConnections* project rooted in the MC^2 and ChemLinks NSF-sponsored curriculum reform projects centered at Berkeley and Beloit.[1] In this project, scenarios impact all components of the learning environment, from lecture to discussion to lab, and therefore it would seem automatic that students would pick up on the context as critical to the success of their learning. For example, their "Do you want fries with that?" module takes the topic of food energy and uses it for the presentation of chemical thermodynamics. Reports by evaluators include evidence that student engagement with the same materials varies depending on the experience and attitude of the graduate assistant instructor (Gutwill–Wise, 2001; Weise et al., 2001; Seymour, 2001). A similar set of issues was recently reported for graduate teaching assistants with inquiry-based instruction, which also highlighted the importance of the instructor's own learning history in teaching reform curricula (Roehrig et al., 2003).

One need not have uncooperative instructors or a poor teaching environment to have trouble with student engagement. Roald Hoffmann and Brian Coppola (1996) provided a rare example of honestly listening to students who rebuff their efforts to be relevant. They teach in two very different contexts. In one case, they taught in a general chemistry classroom; in the other case, they taught in a classroom based on an "organic-first" curriculum (Ege et al, 1997). They tell a story of reading course evaluations (for which they recommended whiskey) and confronting students who do *not* care about the relevance of the class. They share with the reader that "students have not constructed the same understanding that we have…of the subject, its ambience, and its process." And, later, "The students are telling us that you don't have to understand everything in science to learn and use the science."

The appeal of teaching from a relevant context, then, does not always translate into automatic engagement by the students. This invites a more careful consideration of *why* relevant teaching has such promise, and why engagement and learning may sometimes not occur.

Continuity, Interaction, and Openness: Insights from the Learning Theories of Dewey and Mezirow

Relevancy in education has deep roots, one of which is due to the educator John Dewey. Dewey saw himself first as a psychologist and a philosopher. But he sought to make his research and ideas concrete in the context of education. This is certainly the way that he frames his ideas, most particularly in *How We Think* (Dewey, 1997) and *Experience and Education* (Dewey, 1938). Dewey had firmly held ideas about what education was supposed to achieve. It is clear from *How We Think*, for example, that his education was not about teaching students to believe anything. Instead, he started from an observation that they *will* and *do* believe anything ("the mind at every stage of development has its own logic"). Dewey, as a psychologist, was committed to the idea that learners are not blank slates and that learning is something that had to go on in the mind of the particular

---

[1]   It should go without saying that a paper of this length must work from examples. The reader should not misconstrue my selection of particular programs for comment as suggesting that they are uniquely successful or uniquely problematic.

learner. But Dewey, as a philosopher, also saw that teaching involves attaining specific changes in the knowledge and the attitudes of students, to align them with his own view of an ideal society.

Dewey distinguished his method from traditional work by contrasting training with education and by emphasizing the need for the teacher to be responsive. One can train the mind, but that is (for Dewey) about getting the person to act in a certain way, even without understanding. In addition, training fails to enable the student to think about the subject independently. The "training" aim of instruction is behind his analysis of "traditional education." Its goal, as he presents it in *Experience and Education*, includes the following text:

> *Since the subject-matter as well as standards of proper conduct are handed down from the past, the attitude of pupils must, upon the whole, be one of docility, receptivity, and obedience. Books, especially textbooks, are the chief representatives of the lore and wisdom of the past, while teachers are the organs through which pupils are brought into effective connection with the material. Teachers are the agents through which knowledge and skills are communicated and rules of conduct enforced.*

The other option—Dewey's—in working with an immature mind is to educate the students in what he considers a "progressive" method. He suggests that education, in contrast to training, conveys habits of a disciplined mind, wherein "original native endowment [is] turned, through gradual exercise, into effective power."

Dewey was also concerned with *how* the teacher can achieve these goals—that is, where the idea of a relevant curriculum begins. At the center of this method is pedagogy that emphasizes how the experience of the pupil must be the center of instruction. For Dewey, the word "experience" referred to an event. Making an experience meaningful to a child is the key to making an experience education.

Dewey's viewpoint requires two things for an experience to be educationally effective: *continuity* with the pupil's past and future experiences and *interaction* of the experience with "particular individuals at a particular time." Education (instead of training) requires both continuity and interaction. We then come to the question of how to ensure the proper continuity and interaction when a lesson is presented. Dewey had enormous respect for the role of the particular instructor, the one who knows the student and can shape the experience in more personal ways. He recognized the need for the teacher to affect the attitudes of the pupil and, where appropriate, to lead the student in particular paths to achieve the growth and the interaction required.

Dewey's work is almost entirely with children, and his work is cited as part of the rationale for developing and studying "authentic" curricula in K–12 settings. For example, Dewey's ideas gave rise to the concepts of "simulation" and "participation" used in a study of "mutual benefit partnerships," that link a middle school classroom with professionals when learning about technology (Radinsky et al., 2001).

Work by Dewey and others generally does not look at a factor that becomes stronger as a person matures: the fixation of belief in ways that become unchanging. Fixation of belief is present in student's conceptual and factual knowledge (Mulford and Robinson, 2002; Samarapungavan and Robinson, 2001; Nakhleh, 2001; Nakhleh, 1993, Nakhleh, 1992). For example, students misconceive that the properties of a substance are the same as the properties of an atom of the substance, and vice versa.

Another problem in fixation of attitudes involves how students view themselves in relation to learning of a subject. As we mature, we get a better sense of "who we really are," and any attempts to change this will be vulnerable to Barad's charge of "essentializing." This is really an issue of attitude in a way that is well explored in the work of Jack Mezirow as he looked at adult learners (Mezirow, 1991; Wink, 2001b); we can describe this attitude as *openness to learning*. Students who are not open to learn are blocked from continuity, for they cannot see learning as continuous with themselves, however relevant the material may seem to the instructor. Nor are they going to interact with the materials in a manner that gives them meaning, for they don't value the meanings as important to who they are. On the other hand, students who are open to learn can see if something is continuous with themselves, and they are able to look at a topic and think that meaning is something they value. Students will, we hope, come to our classrooms open to learn. If that is the case (and I suspect it is for younger learners), then this factor is not important. But if they are not open—if they have not been emancipated from the blocks put in place by habit—then this must occur first.

Relevance and the Classroom: Practical Components

Continuity, interaction, and openness are simply concepts that still have to be turned into actual teaching practices needed in the relevant classroom. Four components of practice seem important: the **particular student**, the **mediator**, the **materials,** and the **assessments.** All four help ensure that a relevant curriculum doesn't degenerate into the "essentializing" Barad finds in "kitchen physics." When some or all of these components are missing, then barriers between the student and the learning experience arise. Why these components matter and how they can be addressed is summarized in Table 1.

TABLE 1: Practical Components of Relevant Curricula.

Practical Component	Factors for Practice	Addressing the Components in the Classroom
Particular student	If we had one and only one student to work with then the point would be moot. However, we generally do not, and an author of new materials does not even have all the students present. Addressing the needs of each particular student requires attention to language, culture, and the need to reflect on what something means to the individual. In addition, the emotions and attitudes of students are important. Relevant pedagogy has the potential to affect the emotional and attitudinal barriers that may be present.	The center of instruction must link the relevant material and the individual students closely. As such, the object for the particular student is not just mastery of chemistry. It includes mastery of the link between chemistry and the relevant content. Students must also be responsible for using knowledge in context. This responsibility includes alignment with standards established by conventional science and presented by the instructor.
Mediator	On the one hand, the mediator is responsible for keeping students on task and away from making the experience mere entertainment. This, of course, involves holding students to what is most important—learning science. On the other hand, the mediator also looks to help the student keep the relevant scenario in mind and not be thrown out as so much window-dressing. A mediator must also participate in the idea that the relevant material belongs in the content.	The mediator of the instruction must do several things. The mediator's work must include assuring students that the time spent on the context is legitimate to the task of learning the basic science. This, in turn, requires that the mediator frame all or most of the instruction in terms of the relevant content. Ultimately, the mediator should also see that the assessment relates to the context in an authentic way.
Materials	The materials used in a classroom are often generated elsewhere. This is particularly true of textbooks. The materials can address the students, support the mediator, and include assessment tools. If these are done well and are used as such, then the classroom becomes more effective.	The materials must become a valued part of the classroom. This requires that the community of students and instructors recognize some ownership of the materials. Materials must also be level-appropriate so that students are able to engage with them in a meaningful way.
Assessment	Assessment, done well, assigns epistemic responsibility to the students and rewards them appropriately. At the same time, students have a responsibility to be open to learning. This is tied up with the idea of the student as the major or even sole agent of his or her learning.	If students have been brought into ownership of what is taught, then there is an opportunity to make them responsible for demonstrations of what is known, including knowing the context of the scenario. Relevant curricula can include their own inquiry into other aspects of the relevant material and their own decisions about how to present their understandings.

A. The Particular Student: Engagement in order to Learn

As Table 1 implies, Dewey's model of education means that the student is the key person in the classroom. By considering how students engage in general, one can frame the problem of how they engage with relevant material. The "baseline" for relevance is something every student values: what the student should know in order to succeed in a course. Using this baseline to engage students can be done in small and large ways. For example, Metz and Pribyl (1995) described how a series of measurement activities involving the density of a metal brick motivate student learning of significant figures and calculations. A larger example of this vision is the construction of much of the entire course content with students participating as co-designers. An example of this approach is the general education science program described by Middlecamp and Nickel (2000); they reported a course that began with a discussion of what topics the students would like to see addressed. From this list, the instructor chose the particular content.

Particular students are also engaged by curricula that involve role-playing. There is an excellent long-term study that shows the benefits of relevance. It comes from St. Olaf College's analytical chemistry program, which was initially described in 1991 (Walters, 1991a; Walters, 1991b). An extensive recent followup study found benefits to students many years later, even if they did not enter a field that used analytical chemistry (Jackson and Walters, 2000). The benefits for the general student included communication skills and teamwork; for those who said the program positively affected their career choice, real-world experience, impact on self-confidence, and problem solving were also common benefits.

These examples all relate to student learning of traditional chemistry content in a context that ought to have the right amount of continuity and interaction. But there also remains the significant problem of openness. We must get a positive answer to the question *Do the students want and expect to learn in* this *context?* There is good research about how students feel about their own learning (Bennett et al., 2001, Zusho et al., 2003). But there is less on how they feel about the context of their learning, although it is known that openness and engagement are activated when students feel a situation matches their values (Taylor, 2001). Of course, couching a lesson in a particular set of values can also form a barrier to student involvement, especially when these are related to cultural differences in the classroom (Guy, 1999) or to student perception of the content as something alien.

Value-laden curricula are well developed when relevance is built around a shared project that is important to the group. This was demonstrated in two different informal science projects with inner-city gardening projects (Rahm, 2002; Fusco, 2001). Of course, informal settings have the advantage of full participation by the learners, who otherwise would not be there. Bringing values to the forefront of the conventional classroom may seem more difficult. An accessible entry for this lies in discussions of ethics, for which there are many resources, including examples of how to introduce ethics in chemistry courses (Coppola, 2000; Kovac, 1996; Sweeting, 1999). Values-based discussions can also be brought into discussions of different cultural perspectives in the classroom. The efforts in chemical education of "Project Inclusion" (Hayes and Perez, 1997) are a good example of this type of discussion.

B. Materials for Relevant Pedagogy

The materials used for teaching form a second practical element of relevant curricula. Rather than attempt a survey of all that could be done with materials for relevant teaching, I will focus on the particular learning that is possible for materials constructed in a problem-based learning or a gamesmanship/play model.

The method of problem-based learning (PBL) first became prominent in medical schools in the 1980s, and the idea quickly found a home in the basic sciences also. There have been several articles in the chemical education literature reporting the use of PBL, with a useful summary of the early literature already provided by Ram (1999). It is even more prevalent in the biochemistry literature, with a regular PBL feature now in the bimonthly journal *Biochemistry and Molecular Biology Education (BAMBED)* (White, 2002). As Ram (1999) suggests, "PBL problems must be (i) based in compelling, real-world situations, (ii) generate multiple hypotheses, (iii) exercise problem-solving skills and require creative thinking, (iv) require knowledge and skills that satisfy curricular objectives, and (v) be integrated and contain components of more than one discipline."

Many of the examples of PBL in the chemistry curriculum include advanced courses where students have identified a career path. Then, the problem that they are studying is a form of pre-professional training. One paper (Cancilla, 2001) on an environmental chemistry laboratory justifies the work in this way: "…it is essential that environmental chemistry students develop the skills necessary not only to collect and interpret complex data sets but also to communicate their findings in a credible manner in nonscientific forums." In this case, the instructors and students can rely directly on government-issued standards for this work, providing a strong, compelling, and clear external reference point for the entire experience.

PBL is also useful with less advanced students, even complete novices. With efforts to foster engagement, the gamesmanship/play aspect of the experience becomes more pronounced. One example of supporting relevance through gamesmanship or play is in a program to study chemical engineering by relating it to human physiology and drug transport. These programs, titled "Hands-On the Human Body" (Farrell et al., 2002) and "Applied Drug Delivery" (Farrell and Hesketh, 2002), illustrate important aspects of chemical engineering by close studies of physiology from an engineering viewpoint. Not all students are going into fields related to human physiology. But a large fraction of an introductory chemical engineering curriculum is taught in this manner. There is report of strong engagement because the data for many of the experiences are obtained from non-invasive experiments on the students themselves, ranging from tests of expired gases to actual ECGs.

C. Mediation and Classroom Negotiation

Models of teaching where the instructor determines the content and the outcome of the classroom experience are easy to construct and to manage, for all of the items can be in place before the course begins. However, taking students into and through a relevant curriculum requires active negotiation and mediation. For example, Glew (2003) has noted that PBL programs in medical education often face the challenge of improper instructor understanding of the material.

There is now an excellent model for how to train and support mediators: peer-led teaching. Although this doesn't concern relevant teaching itself, the methods developed to train peer leaders are probably useful in the relevant classroom (Gosser et al., 2001; Tien et al., 2002). The training model reported by Gosser and Roth (1998) is representative of this idea. They describe how to take "energetic, smart, and eager" workshop leaders and give them the understanding and skills needed to work in a dynamic, student-centered classroom. The training includes
- Review of the content and practice of the workshop problems by faculty and leaders.
- Instruction and practice in group dynamics and group leadership.
- Discussions about learning styles and intellectual development.
- Practice with new pedagogical methods and review of study tactics.
- Discussions about the impact of race, class, and gender on learning environments.

It is easy to envision how this model can be applied to any new curriculum, not just to the peer-led team learning approach for which it was designed. For example, the peer-led program pays special attention to the interaction mentioned by Dewey, for "…the group leader is central to promoting the mastery, autonomy, and social connections that are integral to the development of intrinsic motivation" (Roth et al., 2001). This model for training has also been applied to PBL for an upper-division biochemistry course (Platt et al., 2003).

D. Relevant Assessment

The fourth element of relevant pedagogy is assessment of student learning. Here, there is a difference that often occurs between different types of "relevant" curricula and materials. On the one hand, if relevance is intended to support student learning of basic content, then we only need to assess students on that basic content. On the other hand, if relevance is also supposed to educate students about the ways chemistry matters to themselves and their world, then assessment should include items about that additional understanding.

The difference here is not just one of coverage, however. The cliché question, *Will it be on the test?* can lead even the most sincere student to reject the relevant context if it is not part of the assessment. This, perhaps, was behind the cynical comments encountered by Hoffmann and Coppola. It certainly was part of the "short-cutting" that we saw in some contexts with our CPLP/WWC materials.

There are three ways in which student understanding of the chemistry in its relevant context can be assessed. They are all summed up in one way: *keep them talking and writing about the context*. The most obvious way to do this is to ask students to pose their answers from a position within the context. So, for example, students can be asked to write a report "as if" they were presenting data to a client. An especially interesting example of this was provided in a report on an upper-division project on environmental analytical chemistry (Cancilla, 2001), where the chemistry students generated reports that were *then* passed on to a coordinated class in environmental law. The audience in this case (law students) then used the chemistry lab reports in further arguments.

Using writing to extend students' thinking matches well the science writing heuristic, originally reported for eighth grade classrooms. The heuristic includes the requirements of a reflection component that is open-ended (Keys et al., 1999). When the method is studied in a general chemistry classroom, student performance in other assessments improves (Rudd et al., 2001), something further confirmed in a study that documented how the improvement in seventh grader's understanding of cell biology after using the heuristic was related to the additional purposeful writing they needed to do (Hand et al., 2004). See Chapter 12.

Another way to continue the relevant context into assessment is to add questions that probe students' understandings in new ways suggested by the context. An example of the former is a link in a *Working with Chemistry* laboratory on carbon dioxide release from decaying biomatter. In the report-writing for this lab, students are sent data on CO_2 levels reported from the Mauna Loa observatory. These data do show the well-known buildup of CO_2 over many decades. But every single summer the CO_2 levels drop, and every single winter they rise. This decrease and increase is due to the earth's natural cycle of vegetative growth and decay with the seasons (biased because more of the earth's land is in the northern hemisphere). In this way, a small context (the decay of leaf matter) leads to a larger context (increases and fluctuations in the amount of global carbon dioxide) during the assessment phase.

Larger assessment methods, spread over an entire course, can also support the integration of learning chemistry and learning about the relevant context. These methods are not developed specifically for relevant teaching. Instead, they emphasize student responsibility for assembling a larger document about their learning over time. One model for this is within a portfolio process (Phelps et al., 1997). Students are asked to create a portfolio with the most representative work from a given time period. While some of the items may be specific assignments, the students must make choices. This means that the work, perhaps originally designed to support learning, is reborn as a student presentation. And that presentation can be a useful place to document student work related to the relevant context. Presentation of student work in this and other ways, such as with poster exams (Mills et al., 2000), are very good tactics to assess students in verbal and public ways.

Virtually all of these particular assessment methods are useful, and they can even be combined, as when Ram (1999) developed a problem-based environmental science laboratory in an integrated sophomore-level lab class. Her assessment of student learning includes four different items besides laboratory reports: self-study logs, self-evaluation logs, oral presentations, and postings on discussion boards. All were credited for the ways they provided additional documentation of student work and learning.

Conclusion: Agents of Reality

The four different practical components that are important to "relevant" teaching (the particular student, the mediator, the materials, and assessment) should, ideally, be integrated in a model of teaching content within a relevant context. There are two such models of the classroom that seem to offer such an environment for such integration.

One model for "relevant" pedagogy is given by the education writer Parker Palmer in his book *The Courage to Teach* (Palmer, 1998). The focus of his book is on the emotional needs of teachers. But he includes an image of the classroom as a "community of truth," developed anew in every classroom by every teacher. Palmer's community of truth model has a lot to say about what can and probably should happen in a relevant classroom. First, the renewal that occurs with every group of students is not automatic; it involves bringing their particulars to learning. For example, some students may not know why saving a patient is important to them. The instructor must negotiate the terms by which they agree that the act of saving a patient must, at the least, give them a better understanding of the relevant chemistry. Second, the conclusions that students reach about the content must be valued; these, after all, are what they will produce first from their relationship to the subject. This makes it clearer that they are interacting with, not just receiving, information. And, finally, this model gives students a role in helping to decide whether something is relevant, supporting their own openness to the material.[2]

The community of truth model for integrating continuity, interaction, and openness in a relevant classroom returns us to Karen Barad. The problems she raises with relevance are part of a larger critique of any view of science or science teaching that enforces a strict boundary of science and culture. Strict boundaries are the reason, she suggests, why one can think that it is possible to teach science "in itself" or to add a bit of culture to make science palatable. In answer to this, she aims to erase these boundaries completely in ourselves and in our students' understandings. As she writes (Barad, 2000), "...what is being described by our theories is not nature, but our participation *within* nature." This allows her to suggest, in her own neologism, that scientists and science students "intra-act" with nature. To teach *this* kind of science requires both content and context with an emphasis on what phenomena mean for students, for science, and for society (and what these three things mean for the phenomena). In her model, then, the question of relevance all but disappears because, she asserts, the only way to know any science is in its relationship with nature, self, and society. That is why her course (for non-majors, admittedly) is constructed from texts about physics, about personal knowledge, and about society.

There is no doubt that Barad's model is radical, and it is indeed based on a philosophy of science that some would find erroneous in its postmodern assertion of the intertwining of nature, apparatus, phenomena, and persons (Haack, 1996). But the holism involved in her view does suggest a viewpoint that will remove some of the boundary issues that cause relevance, content, and students to sometimes fit together poorly.

Suggestions for Further Reading

Chemists interested in teaching in a relevant context have other resources to use besides those discussed in this paper. One example is Dudley Herron's text *The Chemistry Classroom* (Herron, 1996). He includes an entire section on "Things in the Affective Domain," which contains short and helpful discussions on how to motivate students with particular emphasis on values and student interests. The relationship of relevant teaching to chemistry and instructional standards is discussed in the ACS publication *Chemistry in the National Science Education Standards*. Conrad Stanitski (1997) relates the standards to the relevance-rich high school text *ChemCom*, and Mary Virginia Orna (1997) covers the link of the standards to the history and nature of science, both important springboards for relevant teaching. Finally, Bennett and Holman (2002) survey the rationale and some of the research associated with "context-based approaches" in pre-college and college classrooms.

References

American Chemical Society (2001). *Chemistry in the Community (ChemCom)*. New York: W. H. Freeman.

American Chemical Society (2003). *Chemistry in Context, 4th ed.* C. Stanitski (Ed.) New York: McGraw-Hill.

American Chemical Society (2004). *Chemistry*. J. Bell (Ed.) New York: W. H. Freeman.

Barad, K. (2000). Reconceiving Scientific Literacy as Agential Literacy. In R. Reid and S. Traweek., (Ed.) *Doing Science + Culture*. New York: Routledge.

Bennet, J., Rollnick, M., Green, G., and White, M. (2001). The development and use of an instrument to assess students' attitude to the study of chemistry. *International Journal of Science Education*. 23, 833.

---

[2]   It is important to note that valuing students' understandings and conclusions is not the same as validating them. Conclusions about the subject matter must be tested against the "established conclusions" that the instructor carries into the classroom.

Bennett, J., and Holman, J. (2002). Context-based approaches to the teaching of chemistry: What are they and what are their effects.? In *Chemical Education: Towards Research-based Practice*. J. K. Gilbert, O. De Jong, R. Justi, D. F. Treagust, and J. H. Van Driel(Eds.) Dordrecht: Kluwer.

Bransford, J. D., Sherwood, R. S., Hasselbring, T. S., Kinzer, C. K., and Williams, S. M. (1990). Anchored instruction: Why we need it and how technology can help. In D. Nix and R. Spiro (Eds.) *Cognition, education, and multi-media: Exploring ideas in high technology*. Hillsdale, NJ: Erlbaum.

Cancilla, D. A. (2001). Integration of environmental analytical chemistry with environmental law: The development of a problem-based laboratory. *Journal of Chemical Education*. 78, 1652.

Chapman, O. (1993). In National Science Foundation: *Innovation and Change in the Chemistry Curriculum (NSF 94–19)*. Washington, DC: National Science Foundation.

Cognition and Technology Group at Vanderbilt. (March 1993). Anchored Instruction and situated cognition revisited. *Educational Technology*. 52.

Coppola, B. (2000). Targeting entry points for ethics in chemistry teaching and learning. *Journal of Chemical Education*. 77, 1506.

Dewey, J. (1938). *Experience and Education*. New York: Touchstone.

Dewey, J. (1997). *How We Think*. New York: Dover. Originally published 1910.

Ege, S. N., Coppola, B. P., and Lawton, R. G. (1997). The University of Michigan undergraduate chemistry curriculum 1. philosophy, curriculum, and the nature of change. *Journal of Chemical Education*. 74, 74.

Farrell, S., Hesketh, R. P., Hollar, K. A., Savelski, M. J., and Specht, R. (2002). Don't waste your breath. *Proceedings of the 2002 American Society for Engineering Education Annual Conference and Exposition*. Session 1613.

Farrell, S., and Hesketh, R. P. (2002). An introduction to drug delivery for chemical engineers. *Chemical Engineering Education*. 36, No. 3.

Fusco, D. (2001). Creating relevant science through urban planning and gardening. *Journal of Research in Science Teaching*. 38, 860.

Glew, R. H. (2003). The problem with problem-based medical education: Promises not kept. *Biochemistry and Molecular Biology Education*. 31, 52.

Goll, J. G., and Mundinger, S. L. (2003). Teaching chemistry using *From the Earth to the Moon. Journal of Chemical Education*. 80, 292.

Goll, J. G., and Woods, B. J. (1999). Teaching chemistry using the movie *Apollo 13. Journal of Chemical Education*. 76, 506.

Gosser, D. K., Jr., and Roth, V. (1998). The Workshop Chemistry Project: Peer-Led Team Learning. *Journal of Chemical Education*. 75, 185.

Gosser, J. K., Cracolice, M. S., Kampmeier, J. A., Rother, V., Strozak, V. S., and Varma–Nelson, P., Eds. (2001). *Peer-Led Team Learning: A Guidebook*. Upper Saddle River, NJ: Prentice Hall.

Gutwill–Wise, J. P. (2001). The impact of active and context-based learning in introductory chemistry courses: An early evaluation of the modular approach. *Journal of Chemical Education*. 78, 684.

Guy, T. C. (1999). Culture as context for adult education: The need for culturally relevant adult education. *New Directions for Adult and Continuing Education*. 82, 5.

Haack, S. (1996). Science as Social?—Yes and No. In L. H. Nelson. and J. Nelson (Eds.) *Feminism, Science, and the Philosophy of Science*. London: Kluwer.

Hand, B., Hohenshell, L., Prain, V. (2004). Exploring students' responses to conceptual questions when engaged with planned writing experiences: A study with year 10 science students. *Journal of Research in Science Teaching*. 41, 186.

Hayes, J., and Perez, P. (1997). Project Inclusion: Native American plant dyes. *Chemical Heritage*. 15(1).

Herron, J. D. (1996). *The Chemistry Classroom: Formulas for Successful Teaching*. Washington: American Chemical Society.

Hoffmann, R., and Coppola, B. (1996). Some heretical thoughts on what our students are telling us. *Journal of College Science Teaching*. 25, 390.

Jackson, P. T., and Walters, J. P. (2000). Role-playing in analytical chemistry: The alumni speak. *Journal of Chemical Education*. 77, 1019.

Keys, C. W., Hand, B., Prain, V., and Collins, S. (1999). Using the science-writing heuristic as a tool for learning from laboratory investigations in secondary science. *Journal of Research in Science Teaching*. 36, 1065.

Kovac, J. (1996). Scientific ethics in chemical education. *Journal of Chemical Education*. 73, 926.

McCourt, F. (1999). '*Tis*: *A Memoir*. New York: Scribner.

Metz, P. A., and Pribyl, J. R. (1995). Measuring with a purpose: Involving students in the learning process. *Journal of Chemical Education*. 72, 130.

Mezirow, J. (1991). *Transformative Dimensions of Adult Education*. San Francisco: Jossey-Bass.

Middlecamp, C. H., and Nickel, A. L. (2000). Doing science and asking questions: An interactive exercise. *Journal of Chemical Education*. 77, 50.

Mills, P. A., Sweeney, W. V., DeMeo, S., Marino, R., and Clarkson, S. (2000). Using poster sessions as an alternative to written examinations—The poster exam. *Journal of Chemical Education*. 77, 1158.

Mulford, D. R., and Robinson, W. R. (2002). An inventory for alternate conceptions among first-semester general chemistry students. *Journal of Chemical Education*. 79, 739.

Nakhleh, M. B. (2001). Theories or fragments? The debate over learners' naive ideas about science. *Journal of Chemical Education*. 78, 1107.

Nakhleh, M. B. (1993). Are our students conceptual thinkers or algorithmic problem solvers? Identifying conceptual students in general chemistry. *Journal of Chemical Education*. 70, 52.

Nakhleh, M. B. (1992). Why some students don't learn chemistry: Chemical misconceptions. *Journal of Chemical Education*. 69, 191.

Orna, M. V. (1997). The standards and the history and nature of science. In *Chemistry and the National Science Education Standards*. Washington: American Chemical Society. The text parts of this publication are also available from the ACS education website.

Palmer, P. (1998). *The Courage to Teach*. San Francisco: Jossey-Bass.

Phelps, A. J., LaPorte, M. M., and Mahood, A. (1997). Portfolio assessment in high school chemistry: One teacher's guidelines. *Journal of Chemical Education*. 74, 528.

Platt, T., Barber, E., Yoshinaka, A., and Roth, V. An innovative selection and training program for problem-based learning (PBL) workshop leaders in biochemistry. *Biochemistry and Molecular Biology Education*. 31, 132.

Radinsky, J., Bouillion, L., Lento, E. M. and Gomez, L. M. (2001). Mutual benefit partnership: a curricular design for authenticity. *Journal of Curriculum Studies*, 33, 405.

Rahm, J. (2002). Emergent learning opportunities in an inner-city youth gardening program. *Journal of Research in Science Teaching*. 39, 164.

Ram, P. (1999). Problem-based learning in undergraduate education; A sophomore chemistry laboratory. *Journal of Chemical Education*. 76, 1122.

Roehrig, G. H., Luft, J. A., Kurdziel, J. P., and Turner, J. A. (2003). Graduate teaching assistants and inquiry-based instruction: Implications for graduate teaching assistant training. *Journal of Chemical Education*. 80, 1206.

Roth, V., Goldstein, E., and Marcus, G. (2001). *Peer-Led Team Learning: A Handbook for Team Leaders*. Upper Saddle River, NJ: Prentice Hall.

Rudd, J. A. II, Greenbowe, T. J., Hand, B., and Legg, M. J. (2001). Using the science writing heuristic to move toward an inquiry-based laboratory curriculum: An example from physical equilibrium. *Journal of Chemical Education*. 78, 1680.

Russell, A. A., Chapman, O. L., and Wegner, P. A. (1998). Molecular Science: Network-deliverable curricula. *Journal of Chemical Education*. 75, 578.

Samarapungavan, A., and Robinson, W. R. (2001). Implications of cognitive science research for models of the science learner. *Journal of Chemical Education*, 78, 1107.

Seymour, E. (2001). Tracking the processes of change in US undergraduate education in science, mathematics, engineering, and technology. *Science Education*. 86, 79,

Sherwood, R. D., Kinzer, C. K., Bransford, J. D., and Franks, J. J. (1987). Some benefits of creating macro-contextsfor science instruction: Initial findings. *Journal of Research in Science Teaching*. 24, 417.

Staniski, C. (1997). Science in personal and social perspectives and *ChemCom*. In *Chemistry and the National Science Education Standards*. Washington: American Chemical Society. The text parts of this publication are also available from the ACS education website.

Sweeting, L. (1999). Ethics in science for undergraduate students. *Journal of Chemical Education*. 76, 369.

Taylor, E. W. (2001). Transformative Learning theory: A neurobiological perspective of the role of emotions and unconscious ways of knowing. *International Journal of Lifelong Education*. 20, 218.

Tien, L. T., Roth, V., and Kampmeier, J. A. (2002). Implementation of a peer-led team learning instructional approach in an undergraduate organic chemistry course. *Journal of Research in Science Teaching*. 39, 606.

Walters, J. P. (1991a). Role-playing analytical chemistry laboratories. Part 1: Structural and pedagogical ideas. *Analytical Chemistry*, 63, 977A.

Walters, J. P. (1991b). Role-playing analytical chemistry laboratories. Part 3: Experiment objectives and design. *Analytical Chemistry*, 63, 1179A.

Weise, D. J., Pedersen–Gallegos, L., Seymour, E., and Hunter, A. B. (January 2001). The role of the teaching assistant in undergraduate chemistry reform. Paper to the 6th Gordon Conference on Innovations in College Chemistry Teaching, Ventura, CA.

White, H. B. (2002). Commentary: The promise of problem-based learning. *Biochemistry and Molecular Biology Education*. 30, 419.

Wink, D. J. (2001a). Almost like weighing someone's soul: Chemistry in contemporary film. *Journal of Chemical Education*. 78, 481.

Wink, D. J. (2001b). Reconstructing student meaning: A theory of perspective transformation. *Journal of Chemical Education*. 78, 1107.

Wink, D. J., Fetzer Gislason, S., and Ellefson Kuehn, J. (2005). *Working with Chemistry*. 2nd. Edition. New York: W. H. Freeman.

Winthur, A. A. and Volk, T. L. (1994). Comparing achievement of inner-city high school student in traditional versus STS-based chemistry courses. *Journal of Chemical Education*. 71, 501.

Zusho, A., Pintrich, P. R., and Coppola, B. (2003). Skill and will: the role of motivation and cognition in the learning of college chemistry. *International Journal of Science Education*. 25, 1081.

6

Models and Modeling

George M. Bodner
Department of Chemistry
Purdue University

David E. Gardner
Department of Physical Sciences
Lander University

Michael W. Briggs
Department of Chemistry
Purdue University

Abstract

The discussion of models and modeling in this chapter is based, in part, on the results of research on students' understanding of quantum mechanics (Gardner, 2002) and research designed to probe students' ability to visualize the rotation of organic molecules (Briggs, 2004). Our perspective on models and modeling has also been influenced by work in both physics education (Halloun, 1996) and mathematics education (Lesh and Lamon, 1992; Lesh and Doerr, 2003). In this chapter, we argue that the meaning of the term *model* changes as students proceed through the chemistry curriculum, from general chemistry through physical chemistry. We argue that a set of models lies at the core of any scientific theory and that practicing scientists are often actively involved in the process of modeling. We examine attempts to bring models and modeling into mathematics education, and then examine the implications of bringing a models-and-modeling perspective to chemistry courses at various levels.

Biographies

George Bodner is the Arthur E. Kelly Distinguished Professor of Chemistry, Education, and Engineering at Purdue University. He began his academic career as a history/philosophy major at the institution now known as the University at Buffalo. He found, much to his amazement, that chemistry was fun, and he changed his major under the mistaken impression that jobs were easier to find as a chemist. After a mediocre career as an undergraduate (B.S., 1969), he entered graduate school at Indiana University (Ph.D., 1972) where he apparently did well enough as a double major in inorganic and organic chemistry to gain an appointment as a visiting assistant professor at the University of Illinois (1972–1975). Two things became self-evident during his tenure at Illinois: he found that teaching was fun, and he realized that his research could best be described as searching for definitive answers to questions that no one ever asked. When the time came to leave Illinois, he therefore took a job as two-thirds of the chemistry faculty at Stephens College; he lasted for two years (1975–1977) teaching general, organic, inorganic, and biochemistry. He moved to Purdue University in 1977 to take a position in something known as "chemical education." He is the author of more than 100 papers and 45 books or laboratory manuals. His interests include the development of materials to assist undergraduate instruction, research on how students learn, and the history and philosophy of science. Last year, he was selected to receive the Nyholm Medal from the Royal Society of Chemistry, the Pimentel Award in Chemical Education from the American Chemical Society, and the Distinguished Alumni Award from his alma mater, the University at Buffalo.

David E. Gardner attended Carnegie Mellon University and graduated with a Bachelor of Science degree in Physics, with a Chemistry minor, in 1994. From there, he entered Purdue University for graduate work in physical chemistry. After completing an M.S. thesis on the structure of zeolites, he entered the chemical education program, where he completed a Ph.D. dissertation on the learning of quantum mechanics. He is now an assistant professor of physical chemistry at Lander University in Greenwood, South Carolina.

Michael W. Briggs attended the University of Akron before serving as an Army officer in Vietnam. After a 20-year career in the rubber industry, he went back to school to get an M.S. degree at Indiana University of Pennsylvania. After five years at IU–P, he enrolled in the Ph.D. program in chemistry education at Purdue University. Upon graduation from Purdue, he returned to IU–P as a member of the faculty at that institution.

Definitions of *Model* and *Modeling*

The Oxford English Dictionary notes that the term *model* can be used as a noun or adjective, and it means, "a simplified or idealized description or conception of a particular system, situation, or process, often in mathematical terms, that is put forward as a basis for theoretical or empirical understanding, or for calculations, predictions, etc.; a conceptual or mental representation of something." The kinetic molecular theory would be an example of such a model. When we describe Boyle's law or Charles' law to our students, we are talking about these laws in the sense of a model expressed in mathematical terms that can be used to either explain or predict certain properties of a physical system.

The OED also notes, however, that the term *model* can be used as a verb in the following sense: "To devise a (usually mathematical) model or simplified description of (a phenomenon, system, etc.)." For our purposes, we will use the term *modeling* to describe attempts to construct a model of a system. Modeling would occur in a physical chemistry course, for example, when one tried to apply the one-dimensional particle-in-a-box form of the Schrödinger equation to explain why the β-carotene molecule absorbs radiation with a wavelength of 497 nm.

Various attempts have been made to describe the characteristics of a model within the context of science education, including

- A model is a representation of an idea, object, event, process, or system, which concentrates attention on certain aspects of the system (Gilbert, 1997)—thereby facilitating scientific inquiry (Ingham and Gilbert, 1991).
- Mental models represent significant aspects of our physical and social world, and we manipulate elements of these models when we think, plan, and try to explain events in that world (Bower and Morrow, 1990).
- A model relates to a target system or phenomenon with which we have a common experience or set of experiences (Norman, 1997).
- Models are mental entities that people construct with which they reason; all of our knowledge of the world therefore depends on our ability to construct models of it (Johnson-Laird, 1989).

Harrison and Treagust (2000) argued that "modeling is the essence of thinking and working scientifically" and differentiated between analogical models, such as scaled or exaggerated objects, symbols, equations, graphs, diagrams, and maps, on the one hand, and simulations used in model-based thinking, on the other hand. Greca and Moreira (2000) tried to distinguish between *mental models* that exist within the mind of an individual and *physical* or *conceptual models* that are shared among members of a community as follows: "... whereas mental models are internal, personal, idiosyncratic, incomplete, unstable and essentially functional, conceptual models are external representations that are shared by a given community, and have their coherence with the scientific knowledge of that community."

Gilbert (1997) clarified the use of the term model by differentiating between four closely related ideas. A *mental model* is the product of an individual's thought process; an *expressed model* is produced when a mental model is placed in the public domain through action, speech, or writing; a *consensus model* is an expressed model that has been generally accepted among a community of scientists; and a *teaching model* is an expressed model that was specifically developed to help students understand an historical or conceptual model.

An important aspect of teaching models was revealed in interviews conducted by Grosslight et al. (1991) that probed students' understanding of the role of models in teaching and learning. They noted that 7th-grade students often view models as "little copies of real-world objects." Even 11th-grade students "... still fundamentally see models as representations of real-world objects or events and not as representations of ideas about real-world objects or events." Models of covalent compounds, for example, in which Styrofoam balls are held together by pieces of Velcro are likely to be seen by high-school students, for example, as legitimate models of covalent compounds, not representations of the idea of the structure and bonding in these compounds.

Models and Modeling in Mathematics Education

Particular attention has been paid to the use of models and modeling in mathematics by Lesh and co-workers (Lesh et al., 2000; Lesh and Doerr, 2003). Lesh argues that models can be used to describe a system, to think about it, to make sense of it, to explain it, or to make predictions about it. Thus, models can be predictive, interpretative, and/or analytic, not just examples of the system to which they refer. Models are tools that embody the characteristics of phenomena that theory defines as important (Penner et al., 1997). The theory can be as simple as naive conceptions held by a beginner or as complex as carefully studied scientific hypotheses. Because models provide the basis for drawing inferences, they enable new knowledge to be created from former knowledge.

Model-Based Learning

Halloun (1997) has argued that a set of models lies at the core of any scientific theory. Harrison and Treagust (2000) have extended this argument by noting that the construction and evaluation of models is the essence of scientific thinking. If they are right, one might expect that model-based learning would be an explicit part of science courses. But there is abundant evidence to suggest that it is not.

Lesh and co-workers (Lesh et al., 2000; Lesh and Doerr, 2003) have provided a basis for implementing model-based learning in science and mathematics courses in the form of something they call "model-eliciting activities." The products that students produce as a result of a model-eliciting activity go beyond the short answers to narrowly defined questions that have dominated our classrooms for so many years. These products involve the creation of mental models and conceptual tools that can be shared with other students and that can be manipulated, modified, and reused to control, describe, explain, manipulate, predict, or control significant real-world systems (Lesh and Doerr, 2003).

As an illustration of what Lesh and co-workers mean when they talk about narrowly defined classroom questions, consider the following item from one of the National Assessment of Educational Progress exams that asked 8th-grade mathematics students: *If a piece of wood 7'3" long is cut into three equal pieces, how long is each piece?* This was a multiple-choice question, and the accepted answer was: 2'5". Unfortunately, the only place in which this answer could be achieved is the traditional mathematics classroom. In the real world, one might choose to cut the board lengthwise to obtain three pieces that are all 7' 3" long—which was not one of the allowed choices for fairly obvious reasons. Or one might get three pieces that are 2'4¾" or perhaps 2'4⅞", depending on the blade that was used to cut the wood.

Lesh and co-workers argue that traditional textbook word problems are difficult for many students because they require the students to make sense of symbolically described situations. Model-eliciting activities, they argue, are different because they are based on real-life situations for which students have to construct symbolic descriptions. Within the context of mathematics, where model-eliciting activities were first developed, the activity usually involves "mathematizing" a real-world situation. This often involves searching for relationships between objects; sorting or categorizing objects so that they can then be quantified; finding patterns in the behavior of objects; searching for appropriate dimensions in which to express the magnitude of these objects; and so on. Lesh and co-workers note that students given model-eliciting activities often invent, extend, refine, or revise constructs that are more powerful than anybody has dared to try to teach them using traditional methods (Lesh, et al., 1993).

The first step toward bringing a models-and-modeling perspective to your course is relatively easy. If you are teaching general chemistry, you could start by helping your students understand the meaning of the term "law" when it is used in the context of Boyle's Law, or Charles' Law, or Dalton's Law of Partial Pressures. Our experience has shown that many students in introductory courses believe that a "law" is "something that must be obeyed." In fact, this term is used in the sense of a mathematic equation—or "model"—that fits experimental data, more or less, under certain conditions and within certain limitations (Bodner and Pardue, 1995). At various points during the course, you might build on this idea to convey the notion that one of the ways scientists think is in terms of constructing, evaluating, refining, adapting, modifying, and extending models that are based on the experiences with the world in which they live and work. You might also get students to talk to each other, so they may verbalize what it means to use a model such as Boyle's law in a gas law problem, for example.

Lesh and Doerr (2003) describe the traditional view of problem solving as the process of getting from givens to goals when the path is not obvious. They note, however, that problem solving in the traditional classroom is constrained to answering questions using facts and rules that are restricted in ways that are artificial and unrealistic. This is not unique to mathematics, of course. Consider the chemistry class, for example, in which students are asked to answer questions that deal with the bonds between atoms under the artificial and unrealistic constraint that all bonds are presumed to fall somewhere on a continuum upon which ionic and covalent bonds represent the two extremes. They might be able to apply this model to the more or less covalent bonds in the plastic lining on the inside surface of the soft drink cans they bring to class. But they wouldn't be able to apply it to the bonds between the atoms in the aluminum metal from which the can is made because these bonds are neither ionic, nor covalent, nor any combination of these extremes. Or consider the task of classifying compounds as either "more ionic" or "more covalent" on the basis of the traditional rule that predictions of bond type can be based on calculations of the difference between the electronegativities of the atoms that form this bond. The artificial and unrealistic constraints inherent in this rule guarantee an answer to questions of this nature, and they specify a path between the givens and goals of the problem, but they do not necessarily specify the correct answer or the correct path. MgH_2 can be considered to be a source of the Mg^{2+} and H^- ions, whereas CO_2 is a covalent molecule held together by C=O double bonds, for example, in spite of the fact that the difference between the electronegativities of the elements that form these compounds is the same, within experimental error ($\Delta EN = 0.89$). As has been noted elsewhere (Spencer, Bodner, and Rickard, 2005), the correct predictions can be obtained by combining calculations of the average electronegativity of the elements in a bond with calculations of the difference between the electronegativities of these elements along a path that is not inherently rule driven.

Lesh and Doerr (2003) argue that bringing a models-and-modeling approach to problem solving would emphasize "important aspects of real-life problem solving, which involves developing useful ways to *interpret* the nature of givens, goals, possible solution paths, and patterns and regularities beneath the surface of things." They note that the process of getting the answer to real-life problems involves "... 'modeling cycles' in which descriptions, explanations, and predictions are gradually refined and elaborated."

Because these activities were first developed for use in mathematics, let's look at a model-eliciting activity developed for use in a middle-school math class:

- John is constructing a recreation room in his basement. He has put up the walls and put down a floor. He needs to buy baseboard to put along the walls. The room is 21 feet by 28 feet. The baseboards come in 10-foot and 16-foot lengths. How many of each kind should he buy?
- If John wants to have as few seams as possible, how many of each size of baseboard should he buy?
- If John wants to have as little waste as possible, how many of each size should he buy?
- If the 16-foot boards cost $1.25 per foot and the 10-foot baseboards cost $1.10 per foot, how many of each kind should he buy if he wants to spend the least amount of money?
- There is a sale on the 16-foot baseboards. They now cost $0.85 per foot, whereas the 10-foot baseboards still cost $1.10 per foot. How many of each should he buy if he wants to spend the least amount of money?

Like so many others, this model-eliciting activity is based on a "real-world" problem that takes the activity to the student, rather than trying to construct a virtual world in which the student has to come to the instructor. Model-eliciting activities differ from traditional problem-solving activities because they are designed to help students adapt, refine, and modify many of the concepts they already have, and find new ideas to apply to a problem.

Lesh and co-workers (2000) have enunciated six principles upon which the design of model-eliciting activities should be based. These principles involve the following guiding questions:

- The model construction principle: Does the task put the students in a situation where they recognize the need to develop a model for interpreting the givens, goals, and possible solution processes in a complex, problem-solving situation?
- The reality or meaningfulness principle: Could this happen in a real-life situation?
- The self-assessment principle: Does the problem statement clearly indicate appropriate criteria for assessing the usefulness of alternative solutions? Will the students know when they are finished with the problem? Is the purpose clear?
- The construct documentation principle: Will responding to the question require students to reveal explicitly how they are thinking about the situation by revealing how they took into account the givens, goals, and possible solution paths?
- The construct shareability/reusability principle: Is the model that is developed useful only to the person who developed it and applicable only to the particular situation presented in the problem, or does it provide a way of thinking that is shareable, transportable, easily modifiable, and reusable?
- The effective prototype principle: Does the solution provide a useful prototype, metaphor, or "tool" for interpreting other situations?

The fact that model-eliciting activities were first developed in mathematics education does not mean that they are not useful or applicable to chemistry, only that less explicit attention has been paid to the models and modeling perspective by chemists. We have begun the construction of model-eliciting activities for use in chemistry, the first of which deals with the mental rotation of the structure of molecules for use in a sophomore-level organic chemistry course (Briggs, 2002).

Introducing a Models-and-Modeling Perspective

We have already provided an example of how a models-and-modeling perspective might be brought to a general chemistry course by helping students understand the meaning of the term "law" when it is used in the context of the various gas laws. The next step that might be taken in introductory courses would be to recognize that models are almost always used in these courses in the sense of a noun or adjective. They are almost never used in the sense of a verb; in the sense of bringing the construction *and evaluation* of models into the curriculum. Most students walk out of a typical general chemistry course with the feeling that Boyle's law can be used successfully to predict the pressure of a gas from its volume, or vice versa, without any appreciation for the conditions under which the results of this prediction are reasonable.

Instructors might bring a models-and-modeling approach to their organic chemistry courses if they recognize the limitations of discussions of steric effects in the boat conformer of *cis*-1,4-dimethylcyclohexane, for example.

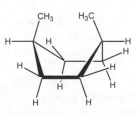

For at least 40 years, organic textbooks have noted the steric repulsion between the two methyl groups in this conformer, often talking about "bowsprit–bowsprit" repulsion (Roberts and Caserio, 1964). It is almost 40 years, however, since the senior author of this chapter built a model of this compound with a plastic model kit, only to find that the hydrogens on one methyl group didn't seem to "touch" the hydrogens on the other. In fact, they didn't seem to even get close to each other.

He now knows that this is a models-and-modeling problem. The mechanical models of organic compounds we assemble from pieces of plastic or steel and the representations of these molecules we build on our personal computers are based on a "hard-sphere" approximation, in which it is assumed that hydrogen atoms are relatively small compared with carbon atoms. If you calculate the distance between the nuclei of the carbon atoms of the two methyl groups in the conformer shown above (3.38 Å), however, you'll find that it is much larger than the sum of traditional estimates of the size of the relevant atoms (≈2.28 Å).

A similar problem arises when one tries to build a model that can explain the steric repulsion in such common examples as the various conformers of *n*-butane.

No matter what model you attempt to build, the distance between the two methyl groups in the eclipsed conformer of this molecule is simply too large to allow for interactions of the magnitude described in introductory organic textbooks. The source of the problem, once again, is the hard-sphere approximation for the size of hydrogen atoms (Tinoco et al., 1995).

The steric effect in either *cis*-1,4-dimethylcyclohexane or *n*-butane can be understood more easily if one thinks about this interaction in terms of a model based on the van der Waals radius of hydrogen, which is more than three times as large as the covalent radius of this atom. Once this is done, the steric effect organic chemists talk about becomes immediately apparent. If one wants to quantitate this interaction, one can think about it in terms of the Lennard-Jones "6–12" potential, namely,

$$\mu = 4\varepsilon \left[\left(\frac{\sigma}{r} \right)^{12} - \left(\frac{\sigma}{r} \right)^{6} \right],$$

where σ reflects the closest distance between the particles undergoing through-space attraction or repulsion, and ε reflects the depth of the potential well. According to recently reported values of the Lennard-Jones parameters (Ben-Amotz et al., 2003), repulsion between the two methyl groups in the boat conformation of cis-1,4-dimethylcyclohexane would occur when the distance between the carbon atoms of these methyl groups is less than 0.373 nm and, as we have seen, the distance between these methyl groups is only 0.338 nm. According to the recently reported values of the LJ 6–12 potential, the magnitude of the repulsion between these methyl groups should be about 7 kJ/mol.

The instructor who brings a models-and-modeling perspective to the sophomore organic chemistry course would be likely to discuss the limitations of the model of the structures of organic molecules based on the traditional hard-sphere approximation that serves as the foundation for our model sets. But, regardless of whether this topic is addressed explicitly in the course, this individual would be ready to explain the apparent dichotomy between the information in the textbook and models of organic compounds when a student raises the question.

Perhaps the best example of a models-and-modeling approach to instruction might involve the first semester of the traditional physical chemistry course (Gardner, 2002). The first lecture typically looks at PVT relationships and reminds students of the ideal gas law:

$$PV = nRT .$$

The second lecture introduces the van der Waals equation, which could be written as

$$(P + \frac{n^2 a}{V^2})(V - nb) = nRT$$

but is often written as

$$(P + \frac{a}{\bar{V}^2})(\bar{V} - b) = RT \,,$$

where $\bar{V}$ is the volume per mole of gas in the sample.

When they discuss the van der Waals model of the behavior of a gas, some instructors compare the results of predictions based on the ideal gas law and the van der Waals equation for a given substance, at a given temperature and volume. Instructors who bring a models-and-modeling perspective to this class wouldn't stop by noting that the predictions of the van der Waals equation are initially smaller and then inevitably larger than those of the ideal gas equation. They would go one step further. They would compare the predictions of these models with the pressure observed experimentally (NIST WebBook). Consider the following results for one mole of CO_2 at 100°C as one gradually decreases the volume of the container.

	Ideal Gas Equation	Van der Waals Equation	Experimental Value
30.6 L:	1.00 atm	0.998 atm	0.998 atm
1 L:	30.6 atm	28.4 atm	28.5 atm
0.200 L:	153.1 atm	104.8 atm	110.5 atm
0.100 L:	306.2 atm	174.9 atm	182.6 atm
0.0500 L:	612.4 atm	2740.7 atm	629.7 atm

When the experimental data are included, one finds that the van der Waals equation is not always a better model of the behavior of the real gas. It provides no advantage at pressures near one atmosphere and it "blows up"—as might be expected—when the volume of the gas starts to resemble the value of the "b" constant in the van der Waals equation.

Instructors who take that extra step at the beginning of their PChem class would set the stage for helping their students recognize that one of the goals of PChem is the construction and testing of models of the behavior of chemical systems, and that one of the goals of the PChem class is to bring "modeling" to the forefront of our discussion of what chemists do.

Implications for Instruction

When one considers the role that models and modeling play in the practice of science, it seems a pity that more attention is not paid to this topic in the science courses we teach. We hope that our examples might help convince the reader that the meaning of the term *model* does, in fact, change as students progress through the chemistry curriculum. In general chemistry and organic chemistry, students are exposed to models as nouns or adjectives. They encounter explicit models, such as the gas laws, the kinetic theory, the collision theory model of chemical reactions, steric effects, and so on. But they are seldom asked to take an active role in *modeling*—in the process of constructing and, more importantly, evaluating models of physical systems. When they encounter physical chemistry, the rules change to some extent. Physical chemistry inherently revolves around both the construction and evaluation of physical models. Unfortunately, many instructors fail to get this message across to their students.

Our colleagues in both mathematics and physics are further along toward an understanding of how to bring a models-and-modeling perspective into their classrooms. We hope that our discussion of model-eliciting activities, however, will induce chemists to develop more examples of their use. It might be noted in passing, however, that there is reason to believe that lecture demonstrations, under the right conditions, can provide a way of bringing the process of modeling into our classrooms. This won't occur when students are passive observers of a "gee-whiz" demonstration done by the instructor. But it might happen under conditions where the demonstration is a discrepant event; where students are actively involved in the demonstration under the conditions that White and Gunstone (1992) refer to as POE: predict, observe, explain; or where the demonstration plays an important role in the first of the four steps in the model of conceptual change proposed by Strike and Posner (1985) which assumes that students must be *dissatisfied* with their present concept, *understand* the concept they have been asked to learn, find the new concept *plausible*, and believe that the new concept is *fruitful* enough to be worth learning.

Work in mathematics by Lesh and co-workers has clearly shown that students subjected to model-eliciting activities can do better in mathematics than anyone expects on the basis of their reaction to traditional instruction. Work in physics by Halloun (1996) has shown very significant improvements in students' ability to handle conceptual questions when instruction is based on a models-and-modeling perspective. Readers interested in additional information on the research literature on the use of models and modeling in chemical education should refer to a recent review by Justi and Gilbert (2002).

Suggested Readings

Gilbert, J. (1993). Models and modeling in science education. Hatfield, UK: The Association for Science Education. This book is still a classic introduction to the role of models and modeling in the teaching and learning of science.

Hestenes, D. (1987). Toward a modeling theory of physics instruction. American Journal of Physics. 55(5), 440–454. [Full-text article available online at http://modeling.la.asu.edu/R&E/Research.html].. Last accessed May 24, 2004. Hestenes was the creator of the Force Concept Inventory. In this paper, he argues that traditional methods for teaching physics are inefficient and concludes that mathematical modeling of the physical world should be the central theme of physics instruction.

Hestenes, D. (1997). Modeling methodology for physics teachers. In E. Redish and J. Rigden (Eds.) The changing role of the physics department in modern universities, American Institute of Physics Part II. 935–957. [Full-text article available online at http://modeling.la.asu.edu/R&E/Research.html]. Last accessed May 24, 2004. Hestenes argues that "scientific practice involves the construction, validation, and application of scientific models, so science instruction should be designed to engage students in making and using models." This article shows how these ideas were incorporated into both ways of teaching physics to students and of training teachers.

Justi, R., and Gilbert, J. (2002). Models and modeling in chemistry education. In Gilbert, J. K., De Jong, O., Justi, R., Treagust, D. F., and Van Driel, J. H. (Eds.) Chemical education: Towards research-based practice. Dordrecht: Kluwer. This chapter presents a review of the limited work that has been done on bringing a modeling perspective to the teaching and learning of chemistry. It addresses the teacher's perspective, the role of chemistry textbooks, and suggestions for "good practice" by textbook authors, teachers, and teacher educators.

Lesh, R., and Doerr, H. M. (2003). Beyond Constructivism: Models and Modeling Perspectives on Mathematics Problem Solving, Learning, and Teaching. Mahwah, NJ: Lawrence Erlbaum. This book represents one of the most extensive examinations of what it means to bring a models-and-modeling perspective to the teaching and learning of mathematics and the training of mathematics teachers. Particular attention is paid to the use of models and modeling in middle-school mathematics and among heavy users of mathematics.

References

Ben–Amotz, D., Gift, andA. D., and Levine, R. D. (2002). Improved corresponding states scaling of equations of state of simple fluids. *J. Chem. Phys.* 117, 4632–4634.

Bodner, G. M., and Pardue, H. L. (1995). *Chemistry: An Experimental Science*, 2nd ed., New York: John Wiley and Sons.

Bower, G. H., and Morrow, D. G. (1990). Mental models in narrative comprehension. *Science.* 247, 44–48.

Briggs, M. W. (2002). *Design of an Instrument to Reveal Subconscious Cognitive Processes During the Mental Rotation of Molecules.* Unpublished M.S. thesis from Purdue University.

Gilbert, J. K. (1997). *Exploring models and modeling in science education and technology education.* Reading, England: The University of Reading.

Greca, I. M., and Moreira, M. A. (2000). Mental models, conceptual models, and modeling. *International Journal of Science Education.* 22(1), 1–11.

Grosslight, L., Unger, C., and Smith, J. E. (1991). Understanding models and their use in science: Conceptions of middle and high school students and experts. *Journal of Research in Science Teaching.* 28, 799–822.

Gardner, D. E. (December 2002). *Learning Quantum Mechanics.* Unpublished Ph.D. dissertation from Purdue University.

Harrison, A. G., and Treagust, D. F. (2000). A typology of school science models. *International Journal of Science Education.* 22(9), 1011–1026.

Johnson–Laird, P. N. (1989). Mental Models. In M. I. Posner (Ed.) *Foundations of cognitive science.* 469–499. Cambridge, MA: MIT Press.

Ingham, A. M., and Gilbert, J. K. (1991). The use of analogue models by students of chemistry at higher educational level. *International Journal of Science Education.* 13(2), 193–202.

Justi, R., and Gilbert, J. (2002). Models and modeling in chemical education. In J. K. Gilbert, O. DeJong, R. Justi, D. F. Treagust, and J. H. Van Driel (Eds.) *Chemical education: Towards research-based practice.* Dodrecht: Kluwer Academic Publishers.

Lesh, R., Hover, M., and Kelly, A. (1993). Equity, assessment and thinking mathematically: Principles for the design of model-eliciting activities. In I. Wirszup and R. Streit (Eds.) *Developments in school mathematics education around the world*: *Vol. 3.* Reston, VA: National Council of Teachers of Mathematics.

Lesh, R., Hoover, M., Hole, B., Kelly, A., and Post, T. (2000). Principles for developing thought-revealing activities for students and teachers. In A. Kelly and R. Lesh (Eds.) *Handbook of research design in mathematics and science education.* 591–645. Mahwah, NJ: Lawrence Erlbaum.

Lesh, R., and Doerr, H. M. (2003). *Beyond constructivism: models and modeling perspectives on mathematics problem solving, learning, and teaching.* Mahwah, NJ: Lawrence Erlbaum.

NIST Chemistry WebBook. Available online at http://webbook.nist.gov/chemistry/.

Norman, O. (1997). Investigating the nature of formal reasoning in chemistry: Testing Lawson's multiple hypothesis theory. *Journal of Research in Science Teaching.* 34(10), 1067–1981.

Penner, D. E., Giles, N. D., Lehrer, R., and Schauble, L. (1997). Building functional models: Designing an elbow. *Journal of Research in Science Teaching*. 34(2), 125–143.

Roberts, J. D., and Caserio, M. C. (1964). *Basic principles of organic chemistry*, New York: W. A. Benjamin.

Spencer, J., Bodner, G. M., and Rickard, L. H. (2005). *Chemistry: Structure and dynamics*, 3rd *Ed.* New York: John Wiley and Sons.

Strike, K. A., and Posner, G. J. (1985). A conceptual change view of learning and understanding. In West, L. H. T. and Pines, A. L. (Eds.) *Cognitive structure and conceptual change.* New York: Academic Press.

Tinoco, I., Jr., Sauer, K., and Wang, J. C. (1995). *Physical chemistry: Principles and applications in biological sciences,* 3rd Ed. Upper Saddle River, NJ: Prentice Hall.

White, R. and Gunstone, R. (1992). *Probing Understanding.* London: The Falmer Press.

7

Enhancing Students' Conceptual Understanding of Chemistry through Integrating the Macroscopic, Particle, and Symbolic Representations of Matter

Dorothy Gabel
Indiana University

Abstract

Students frequently have difficulty learning science because they do not understand how the macroscopic, particulate, and symbolic representations of matter are related. Unfortunately, many chemistry teachers at both the high school and college levels assume that students are familiar with the macroscopic level and teach primarily on the symbolic level. As a result, students try to succeed in chemistry courses by memorization and by solving problems using algorithms rather than by understanding the chemical and physical changes that surround them. Learning theory suggests that, unless students' concepts are linked in long-term memory, they will have an incomplete understanding of a given subject and that students are more likely to learn concepts when learning is active. Active learning provides the time needed for students to confront the alternative conceptions (misconceptions) that they have acquired prior to their formal study of chemistry. This includes their misconceptions about such things as burning, decomposing, melting, and dissolving. Brief descriptions of several studies by chemistry educators will describe some moderate success in helping students link two or more of the macroscopic, particulate and symbolic representations of matter in order to improve their students' understanding of chemistry. Several new programs that provide students the opportunities of making the linkages will also be mentioned.

Biography

Dorothy Gabel, Professor Emeritus of Science Education at Indiana University, received her Ph.D. degree at Purdue University. She was a director of the Teacher Preparation Program at the National Science Foundation in 1987–88, and, for the last 20 years of her service at Indiana University, has supervised and taught a chemistry-based science course for 300 prospective elementary teachers per semester. Dr. Gabel's major research interest is in chemistry problem solving and concept learning. She was a co-author of a Prentice Hall high school chemistry text, the editor of the NSTA monograph on *What Research Says to the Chemistry Teacher about Problem Solving*, and director of the SourceView videotape project on exemplary chemistry teaching. In addition, she was the editor of the *Handbook of Research on Science Teaching and Learning*, and served as president of the National Association for Research in Science Teaching, the School Science and Mathematics Association, and the Hoosier Association of Science Teachers. She has been the recipient of several research awards from the National Association for Research in Science Teaching and a number of service awards from national and state science teachers associations. Her research and teaching publications have appeared in the *Journal of Chemical Education*, the *Journal of Research in Science Teaching*, *Science Education*, and *School Science and Mathematics*. She continues to be actively engaged in encouraging chemistry teachers to use more interactive and inquiry modes of instruction through the use of ConcepTests, Chemical Applications (CAps), and Modeling Activities (MAps).

Why Isn't Chemistry All Students' Favorite Subject?

As an instructor of chemistry, have you ever noticed the expression on someone's face if you introduce yourself as a chemistry teacher at a social event attended by non-chemists? A common reaction that the person gives is some sign of their dislike of chemistry (such as an indication that he or she took chemistry, but never understood it or didn't do well in it) or that you must be a genius (which most chemistry teachers are not)! Even excellent students can be turned off to the study of chemistry because they find it difficult to understand, particularly when they first encounter it.   For example, a young woman who received the highest grade in a large 500-student university introductory chemistry course was asked why she did not take chemistry as an undergraduate. Her response was that she had taken high school chemistry, and although she had gotten an "A," she "didn't understand a thing." So before enrolling in classes at the university, she went to the bookstore and examined textbooks. She found that she could understand the psychology books, the subject interested her, and she decided to major in psychology for her undergraduate degree.   Only later did she decide to study medicine, and hence enrolled in the required chemistry courses!

This student in not alone in her attitude toward chemistry. From informal surveys conducted at Indiana University for the past 20 years at the beginning of a science course for prospective elementary teachers (ninety-nine per cent of whom have had high school chemistry), about 25% of the students reported that they liked chemistry, 25% were neutral and 50 % disliked it. Although a major reason why students' dislike of chemistry may be because they do not do well in a chemistry course, there are also many students who do well, that may not like it.

Ways of Representing and Learning about Matter

Students have difficulty in learning chemistry for a variety of reasons. One difficulty is that chemistry instructors emphasize and teach using abstract symbols. Matter can be studied on three levels: the macro level, the sub-micro level, and the symbolic level (Johnstone, 1991). In chemistry, these are commonly referred to as the macroscopic, particulate, and symbolic levels. The macroscopic level refers to the materials themselves or pictures/illustrations of the materials;, the particulate level refers to atoms, molecules, and ions (or pictures thereof); and the symbolic level refers to symbols for atoms and formulas for molecules, complexes, etc. As Johnstone (1990) has indicated, most chemistry instruction (about 70%) in high school and college chemistry courses takes place at the symbolic level. One reason why students have great difficulty in learning chemistry is because they do not understand the relationship between the symbolic level and the other two levels. They see little or no relationship to the real world with which they are familiar, and they are unable to explain even the simplest chemical reactions in terms of the particle level. Instead, many memorize what is being presented on the symbolic level in terms of chemical equations and mathematical relationships, and solve problems using the factor label method in an algorithmic fashion. If the objective of teaching chemistry is for conceptual understanding, this means that students need to be able to relate the three ways of representing matter.

The two problems given below might be included in a chemistry exam.

1 . A 9.0 mL sample of liquid water is decomposed into hydrogen and oxygen. Calculate the total volume of gas produced at STP if both gases are collected in the same container. Show your work and explain your answer.

A   9.0mL   B   11.2 mL   C   9.0L   D   11.2L

E   16.8L   F None of the above.

2.   After water is decomposed by electrolysis (passing an electric current through it), the oxygen gas and hydrogen gas produced are collected in the same container. Which of the pictures given below represents a small sample of the gaseous mixture? (Each circle represents an atom.) Explain your choice.

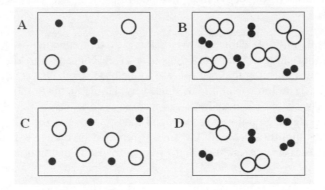

Students will most likely solve the first problem in a routine manner using the factor label method. It requires students to know the formula for water, to be able to write a balanced equation for the decomposition of water, to be able to interpret the equation in terms of moles, to know that hydrogen and oxygen are ideal gases, to know that the volume of an ideal gas at STP is 22.4 liters, and to know that the density of water is 1 g/mL.

The second problem also requires students know the formula for water, to be able to write an equation for the decomposition of water, to know that the equation can be interpreted on the molecular level, and to know that the ratio of hydrogen to oxygen molecules is 2 to 1. It also requires that students realize that hydrogen and oxygen are diatomic gases, and that hydrogen atoms are smaller than oxygen atoms.

The two problems test two different things. Both involve symbols and mathematical reasoning. The second requires a particle interpretation of what occurs, whereas the first does not require such an interpretation. Both are a part of chemistry. For a comprehensive understanding of chemistry, students need to be able to answer both questions correctly.[1]

An alternative approach to teaching chemistry concepts is to have students observe and describe a familiar chemical reaction in the lecture or the lab (such as the burning of charcoal) on the macroscopic level, then to ask them to provide an explanation on the particle (molecular) level in terms of what has occurred, and finally to have them represent the particle level symbolically using a balanced equation. Then present the student with a stoichiomentry problem. This rarely occurs in instruction.

Another complicating factor of why students find chemistry particularly difficult is due to their current understanding of how the world works. Many students have alternative conceptions or misconceptions that are firmly embedded in their long-term memory. A current explanation of how individuals learn based on the information-processing model has been described by Johnstone (1997). Stated briefly, after information is obtained by the senses, some passes into the limited space in short-term/working memory where it is prepared for storage in long-term memory. The storage may be in the form of branched networks (similar to a concept map) or fragments. Because the senses sometimes deceive us, or incorrect information is actually taught, memorized, and/or interpreted differently, individuals have networks that contain alternative conceptions, which are frequently referred to as misconceptions. Once these alternative conceptions are embedded in long-term memory, they are very difficult to replace with concepts generally accepted as common scientific belief. This has important implications for teaching chemistry. For a more complete discussion of the use of the term "alternative conceptions" versus "misconceptions," see Wandersee, Mintzes, and Novak (1994).

---

[1] You might want to administer these two questions (or similar ones) to your students at various points in your chemistry course. I would be interested in the results. If reader response is sufficient, it would make a nice report for JCE!

The existence of these alternative explanations of how the world works can be directly addressed in instruction using the social constructivist model described by Driver and Odham (1986) and Krajcik (1991). Basically, the model consists of students describing their own understanding and restructuring it through exchange with other students and teachers. One method that has become increasingly popular in the teaching of chemistry in the past five years is through the use of ConcepTests, a method of instruction originally proposed by Mazur (1995) for the teaching of physics.

ConcepTests are conceptual, multiple-choice items that are used in instruction for the single purpose of informing the instructor about students' understanding of a given concept. In Mazur's introductory physics class, students generally view a question that is displayed to the class, they answer it, and the individual answer is conveyed to the instructor. The information is then used by the instructor to modify the lecture to take into account students' incorrect answers. ConcepTests can also be used in other ways. Small groups of students sitting near one another may agree on a group answer, and then someone in the group is asked to explain their answer to the class. Groups that disagree with the answer reported give their explanation, and the instructor adds additional insights as necessary. A question that I frequently use in instruction to begin a unit on chemical reactions is as follows:

Which of the following represents the burning process?
A Bread turning black when overheated in the toaster.
B Charcoal glowing on the picnic grill.
C A firefly glowing in the dark.
D More than one of the above.

In fact, I frequently use this question at the beginning of lectures to chemists on the conceptual teaching of chemistry. It creates quite a discussion because the correct answer is B. The general population uses the word " burning" in a variety of ways. Chemists usually define it as a combination reaction in which light and heat are emitted. The "bread turning black" hasn't burned yet, it has just decomposed, and hopefully someone will pull the cord out of the electric socket before it actually burns!

The paperback book *Chemistry ConcepTests: A Pathway to Interactive Classrooms*, by Landis, Ellis, Lisensky, Lorenz, Meeker, and Wamser (2001), is a good resource for more information. The *Journal of Chemical Education* also houses a website (http://www.jce.divched.org/JCEDLib/ QBank/collection/ConcepTests/) containing chemistry ConcepTest questions to which any chemistry teacher can submit and use items, although this does not mean that all of the items on the website are conceptual!

The use of ConcepTests and other social-constructivist models of learning enables a student to construct new meaning by modifying the existing networks in long-term memory. Hence, the student's knowledge becomes more in accordance with an acceptable scientific view.

The Complexity of Chemistry

Chemistry is a very complex subject. The world around us is observed on the macroscopic level. Explanations for these observations can occur on the macroscopic level, the particle level, and the symbolic level. An example of each level of explanation of a candle burning under a jar is given below.

Observations: A candle is lit using a match. A jar is placed over the burning candle. The candle burns for a few minutes and finally goes out. During the process, the wax turns to a liquid, the wick gets black but its tip is red, the jar gets warm, the candle gets smaller, and the flame goes out. The inside of the jar is wiped with a piece of blue cobalt chloride paper and it turns pink. Limewater is poured into the jar, the jar is shaken, and the limewater becomes cloudy. [Many more observations of just the candle burning could be made as indicated in Appendix 1 of the original ChemStudy text, *Chemistry, An Experimental Science* (1963) that lists 53 observations.]

Macroscopic Explanation: For ordinary combustion or burning to occur, oxygen must be present. The wax liquefies in the bowl of the candle due to the heat produced by the wick, turns to a vapor, and then burns as shown by producing light and heat. This continues until most of the oxygen is used up. The tip of the wick is also burning because it produces light and heat. The side of the wick is black, so it probably has decomposed to carbon. The blue cobalt chloride paper turns pink, so water must have been present. (This is, indeed, a test for the presence of water). Oxygen must be present because the lime-water turns cloudy. (This is the test for carbon dioxide).

Particulate Explanation: An explanation on the particle level can be given in terms of kinetic molecular theory. The heat from a match causes the solid to melt, and the particles are more disorganized in the melted state and can move from place to place. Additional heat causes particles to change to a vapor where they occupy about 1000 times more space and allows the paraffin molecules to mix with the oxygen from the air. Particles of oxygen gas come in contact with the wax molecules; in some way, the hydrogen in the wax reacts with the oxygen to form water vapor. The carbon component of the wax reacts with the oxygen to form carbon dioxide gas. The reaction could be represented using three-dimensional models (molecular or Play-Doh), two-dimensional paper and pencil particle pictures, or computer simulations.

Symbolic Explanation: One can give an explanation on the symbolic level by writing a balanced chemical equation: $C_{31}H_{64}$ (s) + 94 O_2 (g) → $31CO_2$ + $32H_2O$.
(Note: Candle wax is a mixture of alkanes and esters. Only this alkane in bees wax is shown above.)

This equation can be interpreted on the molecular level or in terms of moles:

1 molecule of solid paraffin combines with 94 molecules of gaseous oxygen to produce 31 molecules of carbon dioxide gas and 32 molecules of water vapor.

or

1 mole of solid paraffin combines with 94 moles of gaseous oxygen to produce 31 moles of carbon dioxide gas and 32 moles of moles of water vapor.

Another example related to everyday life pertains to the dependence of the boiling point of water on the atmospheric pressure.

Observation: An egg placed in water and boiled in New York takes 3 minutes to become hard, whereas an egg placed in boiling water in Denver is still soft after 3 minutes. Why does it take longer to hard-boil an egg in Denver than in New York City?

Macroscopic Explanation: The boiling point is dependent on the atmospheric pressure. Because New York City is at sea level, the atmospheric pressure is generally about 1 atm. Denver is about a mile above sea level, and hence its atmospheric pressure is less. This results in the boiling temperature of water to be less than 100ºC in Denver; hence, the egg must be boiled for a longer time for the chemical reactions that produce hardening to occur. The higher the temperature, the faster the chemical reaction.

Particle Explanation: When water is heated, the molecules move more quickly, as indicated by an increase in temperature. Some particles move away from others and form small packets of gas in the liquid. As the temperature increases, these tiny bubbles rise to the surface; when boiling begins, they rapidly escape into the atmosphere. The temperature at which they do so is called the boiling point. Because the atmospheric pressure is less at higher altitudes, there is less resistance from the air molecules to prevent them from escaping; hence, they have sufficient energy to do so at a lower temperature. The egg cooks longer at a lower temperature because the chemical reactions in an egg arc temperature dependent. Because temperature is an indication of the speed at which molecules are moving, there will be fewer molecules that have sufficient energy to react until more heat energy is added to the container; this will increase the time of cooking.

Symbolic Explanation: Water changing from a liquid to a gas can be represented by the equation :

H_2O (l) → H_2O (g).

Unfortunately, I do not have an equation for the chemical reaction of hardening an egg! Included in the symbolic level are also mathematical symbols and equations. It is possible for students to use the Clausius–Capeyron equation (included in some introductory college chemistry textbooks) without students ever understanding the particulate explanation. One must ask, which is more important?

Each of these explanations gives different information about a burning candle or why eggs cook faster at lower elevations. If only the symbolic explanation is given, the information may go into short-term memory, but may not be stored in long-term memory—or if it is, it may be stored as an isolated fragment. An expert in chemistry, such as the instructor, would have linkages between the three different representations resulting in a good understanding of burning and reaction time. Novices in the study of chemistry may understand the macroscopic explanation if it is presented to them in class, and linkages to things they have observed might be made in long-term memory. The point is, by starting to teach chemistry primarily at the symbolic level, the subject matter is less related to things with which students are familiar. The result is that students memorize what is meant to be an explanation to pass the course. This approach dampens students' enthusiasm for studying chemistry, and makes chemistry learning boring for many students.

Chemistry instruction at both the high school and college levels frequently fails to take into account that the particle model of matter (including both the kinetic molecular theory and atomic theory) provide the explanations for both matter and its interactions. It appears to be taught as an end in itself. Whether this occurs because chemistry instructors think that these topics can be learned in isolation from macroscopic observations, that perhaps that students have already had these macroscopic observations, or that there is insufficient time is unknown. Whether changes in how chemistry is taught at the high school level will occur with the implementation of the National Science Education Standards (1995) and its emphasis on inquiry and explanations is also not known. This will probably not occur until chemistry instruction at the high school and college freshman level includes more instruction at the macroscopic and particle levels and less at the symbolic and theoretical levels.

From my work as the director of the production of standardized achievement tests (Core40) for each of the sciences in Indiana to be used to determine whether students meet the National and Indiana Science Education Standards, I found chemistry teachers and university professors quite resistant to change. For example, it was difficult to convince chemistry instructors that the conceptual understanding of density (not just substituting numbers in the density formula) was more important than the memorization of electron configurations, and that high school chemistry is the not just "a prep course"" for college chemistry.

The latest edition of "ChemCom," *Chemistry in the Community* (2002), has a more realistic approach to what should be taught at the high school level. When this textbook becomes more widely adopted, perhaps students will be more conceptually prepared for college chemistry. The text contains numerous pictures of the particle nature of matter and frequently relates the pictures to the symbolic representation of the particles. In addition, chemistry is made relevant by linking it to the macroscopic level through real-world applications.

At the college level, the ACS examinations and, in particular, the conceptual general chemistry examination now include particle nature of matter problems, and most current general chemistry textbooks also contain more examples, explanations, and problems involving particles than in the past. In fact, some have an emphasis on those aspects. For example, see Silberberg (1995), Olmsted and Williams (2000), and McMurray and Fay (2004).

Over the past 25 years, many studies have been conducted on students' understanding of chemistry on the macroscopic, symbolic, and particle levels (see Nakleh, 1992), but only a few studies have been reported that show that students who studied chemistry at all three levels understood chemistry better or did better on achievement exams. Most studies only identify students' misconceptions or alternative explanations. Once students have a science misconception or alternative explanation, it is persistent and difficult to dispel unless it is directly addressed because it is embedded in long-term memory. For example, when simple words that are commonly used in everyday language in a non-scientific manner are used in chemistry, such as "dissolving" or "melting," students do not realize that they are not using the word as it is defined in chemistry. (You may wish to consult a dictionary for the definition of "melting" and "dissolving" to see why the problem persists!) Hence, when given questions on all three levels about the same concept, students do not score consistently well on all three ways of understanding a given concept. An example of questions on the macroscopic, particulate, and symbolic level on "melting" is given below:

Macroscopic Question:
Which of the following represents melting?
A A few grains of salt placed in water seemingly disappears.
B Chocolate placed in the sun on a hot day turns to liquid.
C Sugar placed on your tongue produces a sweet taste.
D Two of the above

Particulate Question:
Which of the following represents melting?

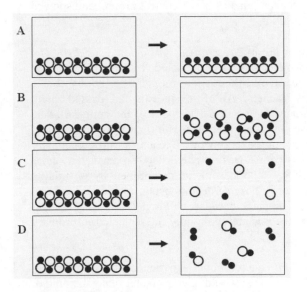

Symbolic Question:
Which of the following represents melting?
A $C_{12}H_{22}O_{11}(s) \rightarrow 12C(s) + 11\ H_2O(g)$
B $C_{12}H_{22}O_{11}(s) \rightarrow C_{12}H_{22}O_{11}(l)$
C $C_{12}H_{22}O_{11}(s) \rightarrow C_{12}H_{22}O_{11}(aq) + H_2O(l)$
D Two of the above

Unless students understand the concept on all three levels, they really do not fully understand the concept. Hence, it is important for chemistry instructors to be aware of these misconceptions because they can be directly addressed and sometimes modified through the use of ConcepTests as described earlier.

An excellent resource that identifies chemistry science misconceptions is *Making Sense of Secondary Science*, by Driver, Squires, Rushmore, and Wood-Robinson (1994). For more information on misconceptions on the particulate nature of matter, see Harrison and Treagust (2002). The next section describes a study of students' understanding of common chemistry processes on the macroscopic, particulate, and symbolic levels.

Studies on Modifying Students' Understanding of Chemistry on the Macroscopic, Particulate, and Symbolic Levels

As indicated earlier, over the past 25 years, only a few studies have been reported that examine students' changes in understanding chemistry at the macroscopic, particle, and symbolic levels as the result of integrating them in instruction. Most studies only identify students' alternative conceptions. A brief description is given below of several studies that have been conducted that have relevance for teaching introductory chemistry courses at the secondary or college levels.

One of the more comprehensive studies in examining the effectiveness of integrating the macroscopic, particulate, and symbolic representations of matter was conducted with high school students and their teachers (Bunce and Gabel, 2002). Ten teachers who had previously attended a summer workshop on the teaching of chemistry conducted the study in their classrooms after collectively preparing three instructional two-week chemistry units and creating test questions on the macroscopic, particle, and symbolic levels. Each teacher randomly assigned one of his/her classes to the control group and another to the treatment group. The study showed that students who were enrolled in the treatment group and made particle pictures scored higher on the particulate questions on the unit exams, but there was no significant difference between students' scores on the macroscopic and symbolic levels. On the delayed posttest given at the end of the semester, there were no significant overall differences at all between the treatment and the control groups. However, an interesting finding of the study was that the visual, particle, and symbolic approach was more effective for females than for males. Female students in the experimental group had significantly higher scores on the particulate and symbolic questions both on the immediate and delayed posttests than females in the control group, and showed a significant gain in achievement over males on the particle questions.

A *Study of College Students' Understanding of Chemistry on the Macroscopic, Particulate, and Symbolic Representations of Matter after Making Play-Doh Models* of molecules was conducted by Gabel, Yang, and Hitt (2003). The students involved in the study were enrolled in an Introduction to Scientific Inquiry course as the first of four science courses for prospective elementary teachers. All of the students had passed a high school chemistry course as a pre-requisite to entrance to Indiana University. The focus of the course was on helping students understand scientific inquiry by learning the skills involved in inquiry, such as observation, prediction, controlling variables, testing hypotheses, and conducting experiments. Because so little chemistry was included in their physical science course, the focus of this course was on chemistry—in particular, those chemistry concepts that are frequently taught at the elementary school level, as determined by examining elementary school science textbooks. These included six concepts: burning/decomposing, melting/dissolving, and chemical/physical change. The concepts were arranged in pairs because children (and many adults) frequently do not distinguish differences within a pair, and the words are commonly interchanged in ordinary usage.

The purpose of this study was to determine whether students who had additional experiences in making three-dimensional models of chemicals undergoing change using the Play-Doh would have a greater understanding of the aforementioned chemistry concepts than students who made two-dimensional paper–and-pencil diagrams. An additional reason for conducting the study was to determine which representation (macroscopic, particulate, or symbolic) and which concept-pairs students found to be most difficult.

Instructors (6) in the comparison group (8 sections of 24 students) had their students make two- dimensional particle pictures in their classes on eight occasions. Instructors of the Play-Doh classes (3 instructors and sections) had students construct Play-Doh models that illustrated certain chemical and physical processes seven times throughout the semester in addition to the drawing of two dimensional pictures. The instrument used for the achievement comparison between the Play-Doh and the control group consisted of 50 questions.

Results indicated that the overall score on the exam for the treatment group was 30.5, whereas the control group's score was 28.5. The difference in scores was significant at the 0.02 level, which indicated the usefulness of using Play-Doh in the instruction. However, the overall score of about 60% correct on the 50-item test was disappointing. The last six items of the achievement test consisted of multiple-choice questions related to the three concept pairs: burning/decomposition, chemical/physical change, and melting/dissolving. Nine questions (three on each concept pair) were included in different versions of the chemistry achievement test so that for each pair, a macroscopic, particulate, and symbolic question was included. Comparisons made on the type of questions indicate that students answered the macroscopic questions correctly more frequently than the particle questions, and that the symbolic questions were more difficult than both the macroscopic and particle questions. In terms of concept pairs, the most difficult pair was melting/dissolving, and the least difficult was chemical/ physical change. Sample questions on each of the three levels of representing "melting" are given in the previous section of this paper.

Results also confirmed the view that understanding chemistry on the symbolic level is more difficult than the other two levels. The difference between students using the Play-Doh models may be due to the increased interaction among students about the concepts that the use of Play-Doh appears to stimulate in the classroom. However, differences may also be due to the difficulty of individual items on the test due to the nature of the representations themselves.

A more recent study by Gabel, Wozniewski, and Cardellini (2004, in preparation) compares the pre–post understanding of chemistry concepts (burning/decomposition, chemical/physical change, and melting/dissolving) on the macroscopic, particulate, and symbolic levels of prospective elementary teachers with college chemistry students in the United States and in Italy. Preliminary findings indicate that there is no significant change from pretest to posttest scores for chemistry students in the United States and Italy (both of whom scored the same). However, through instruction using Play-Doh models, the scores of prospective elementary teachers on the understanding of the same chemistry concepts indicated above changed significantly. The preservice teachers' post-test scores were equivalent to the pretest scores of students who were enrolled in the first or second semester of college chemistry. Since the overall pretest to posttest gain for chemistry students was only 1 point out of a possible 18, this finding shows that traditional introductory chemistry courses do little to improve students' conceptions of chemistry that are generally included in science textbooks at the elementary or middle school level.

Three studies that use computer animations of molecules undergoing changes have had positive effects on the understanding of chemistry. A study by Dori and Hameiri (2003) indicates that by making high school students enrolled in chemistry classes aware of the three levels at which chemistry can be understood using a Multi-Dimensional Analysis System in teaching problem solving, students' success in solving chemistry problems improved. Dori and Hameri used three transformations in the problem-solving process: symbol to macro, symbol to micro, and symbol to process. Students of lower mathematics ability showed a greater increase in problem-solving scores than higher ability students showed, and results were independent of gender. Perhaps if these changes in teaching chemistry were more widespread and were included in chemistry instruction at the college level, more students would be successful problem-solvers.

At the college level, a study by Russell, Kozma, Jones, Wykoff, Marx, and Davis (1997) provides more evidence for improving chemistry understanding by relating the three representations of matter. In the 4M:CHEM program that they created, a computer split-screen design is used; it has four windows that simultaneously show videos of real-life experiments, molecular-level animations of these experiments, symbolic representations, and graphs or diagrams of macroscopic properties and structures. The four windows can be shown individually or in any combination, and when multiple windows are activated, actions in each are synchronized. An examination of the effectiveness of using the macroscopic, particle, and symbolic levels with the module on gaseous equilibrium developed as described above provides evidence for using this approach in instruction. In a study of 295 students enrolled in an introductory chemistry course at the University of Michigan, students' scores increased significantly (p=0.0001) from the pretest given at the beginning of the semester to the posttest given during the last class session after using the 4M:CHEM unit on chemical equilibrium. Also significant (p=0.0001) was the decrease in students' statements of misconceptions at the end of the course. Unfortunately, no comparison was made between students who had used the 4M:CHEM program with those who had not.

An earlier study by Williamson and Abraham (1995) made use of computer animations for other topics in chemistry instruction and produced results similar to those of the aforementioned study. This study, however, made comparisons between students who used the animations with those who did not. In the study, the chemistry educators sought to determine whether computer animation of the particulate nature of matter would enhance the understanding of two units: Solids, Liquids, and Gases; and Reaction Chemistry. They also examined whether students' attitudes (as measured by the BAR test) would be enhanced and whether conceptual understanding would vary according to students' reasoning ability as measured by the TOLT test. Students who participated in the study were from two sections of the first-semester general chemistry course, both taught by the same instructor and both of which had a total enrollment of 400 students. One lecture section was randomly selected as the experimental group that used animations in the lectures and the other section served as the control group that had no animations.

In the experimental group, some discussion sections used additional animations whereas the other discussion sections did not. This resulted in three groups that could be compared. The study used two- to three-minute computer animations in eight areas in teaching a unit on Gases, Liquids, and Solids during six lectures and corresponding discussion sections. Topics included pressure, temperature, diffusion or mixing, ideal gases, phase transitions, liquid vapor equilibrium, intermolecular forces, and London forces. Five animations were used in four lectures for the unit on Reactions. These included the following topics: solutions, ionic solutions, precipitation, temperature effect, and redox. Conceptual understanding was measured using the Particulate Nature of Matter Evaluation Test (PNMET).

One-way analyses of variance were used to determine the effectiveness of using the animations for both the Gases, Liquids, and Solids unit and the Reactions unit. The general findings were that conceptual understanding was significantly incrcascd for students who viewed the animated sequences depicting particulate behavior. These students had fewer misconceptions and also held a more particulate view of matter as determined by their use of conservation of particles in the drawings and fewer "continuous matter" representations of matter.

From the above studies, it appears that computer animations of particles is effective in providing explanations for what can be sensed on the macroscopic level when matter undergoes change. Whether the use of computer animations is more effective than using three-dimensional Play-Doh or other models has yet to be examined. Perhaps different types of models are more or less suitable for different chemistry concepts. The use of a combination of models with students examining the limitations of each may also be an effective way to produce conceptual understanding.

Recommendations and Conclusions

As indicated earlier, chemistry is a very complex discipline, and understanding it on the symbolic level is most difficult even when students have had experiences on the macroscopic and particle levels. One might hope that, as science teachers at all levels more fully implement the National Science Education Standards with the emphasis on inquiry that requires students to give explanations for phenomena they observe, students will memorize less and understand more. This realignment of the curriculum with the Standards will require more class time, and the shift of emphasis will require teachers to present fewer concepts in a sequence, such as those provided as maps in the *Atlas of Science Literacy* (2001), developed by the American Association of the Advancement of Science. Students at the elementary school level need to have a macroscopic understanding of basic chemical and physical processes and changes that matter undergoes. Explanations at the particle level are appropriate at the middle-school level for physical changes and at the high-school level for chemical changes, as well as representing these changes symbolically. Needless to say, students at the college level should understand and be able to make transitions between the three modes of representing matter. In addition, they need to be able to use mathematical equations (also symbolic) in solving chemistry problems.

Hopefully, as college chemistry instructors become more aware of the need to understand and to relate the macroscopic, particulate, and symbolic levels of representing matter, they will be more explicit in chemistry instruction, and college students' understanding of chemistry will increase. Perhaps this increase in understanding will result in less blind memorization on the part of students, and they might even develop a more favorable view of chemistry.

Changes in chemistry courses are occurring in some institutions. Instructors are focusing on fewer concepts, relating chemistry to the real world, requiring students to explain chemistry in terms of the particulate level, and then requiring students to represent their macroscopic observations and particulate explanations using symbols. It is only when students use these techniques that they might have a coherent understanding of the chemistry concepts anchored in their long-term memory.

Suggested Readings

Gabel, D. L.; Samuel, K. V.; Hunn, D. J. (1987) *Journal of Chemical Education*, *64*, 695-697.

Johnstone, A. H. (1997). Chemistry teaching—Science or alchemy? *Journal of Chemical Education*. 74. 262.

References

American Association for the Advancement of Science (2001). *Atlas of Scientific Literacy.* Washington, DC: AAAS and NSTA.

American Chemical Society (2002). *Chemistry in the Community (4th Ed.)* NY: W.H. Freeman.

Asimov, A. (1965). *A Short History of Science.* NY: Doubleday.

Bunce, D. M. and Gabel, D. (2003). Differential Effects on the Achievement of Males and Females of Teaching the Particulate Nature of Matter. *Journal of Research in Science Teaching.* 39, 911.

Burke, B. A., Greenbowe, T. J., Lewis, E., and Peace, G. E. (2002). *Journal of Chemical Education.* 79, 699.

Chemistry: An Experimental Science. (1963). G.C. Pimental (Ed.) CA: W. H. Freeman.

Dori, Y. J. and Hameiri, M. (2003). Multidimensional Analysis System for Quantitative Chemistry Problems: Symbol, Macro, Micro, and Process Aspects. *Journal of Research in Science Teaching.* 40, 278.

Driver, R. and Odham, V. (1986). A constructivist approach to curriculum development in science. *Studies in Science Education.* 13, 105.

Driver, R., Squires, A., Rushworth, P. and Wood–Robinson, V. (1994). *Making Sense of Secondary Science.* London: RoutledgeFalmer.

Gabel, D. (1999). Improving Teaching and Learning through Chemistry Education Research: A Look to the Future. *Journal of Chemical Education.* 76, 548.

Gabel, D. L., Hitt, A. M., and Yang, L. (March 2003). Changing prospective elementary teachers' understanding of the macroscopic, particulate, and symbolic representations of matter using Play-Doh models. Paper presented at the annual meeting of the National Association for Research in Science Teaching, Philadelphia, PA.

Gabel, D. L., Wozniewski, L., and Cardellini, A., (in preparation). A Comparative Study of College Students' Understanding of Everyday Chemistry Concepts.

Harrison, A. G. and Treagust, D. F. (2002). The particulate nature of matter: Challenges in understanding the submicroscopic world. In J.K. Gilbert et al. (Eds.) *Chemical Education: Towards Research-based Practice.* London: Kluwer.

Johnstone, A. H. (September 1990). Fashion, fads, and facts in chemistry education. Paper presented at the American Chemical Society meeting, Washington, DC.

Johnstone, A. H. (1991). Why is science difficult to learn? Things are seldom what they seem. *Journal of Research in Science Teaching.* 40, 278.

Johnstone, A. H. (1997). Chemistry teaching—Science or alchemy? *Journal of Chemical Education.* 74. 262.

Justi, R. and Gilbert, J. (2002). Models and modeling in chemical education. In J.K. Gilbert et al. (Eds.) *Chemical Education: Towards Research-based Practice.* London: Kluwer.

Krajcik, J. S. (1991). Developing students' understanding of chemical concepts. In S.M. Glynn and R.H. Yeany and B.K. Britton (Eds.) *The Psychology of Learning Science.* Hillsdale, NJ: Lawrence Erlbaum.

Landis, C. R., Ellis, A. B., Lisensky, G. C., Lorenz, J. K., Meeker K., and Wamser, C. C. (2001). *Chemistry ConcepTests: A Pathway to Interactive Classrooms.* Upper Saddle River, NJ: Prentice Hall.

Mazur, E. (1997). *Peer Instruction.* Upper Saddle River, NJ: Prentice Hall.

McMurry, J. and Fay, R. C. (2004). *Chemistry.* Upper Saddle River, NJ: Prentice Hall.

Nakhleh, M. B. (1992). Why some students don't learn chemistry: Chemical misconceptions. *Journal of Chemical Education.* 69, 191.

National Research Council (1995). *National Science Education Standards.* Washington, DC: National Academy Press.

Olmstead, J. and Williams, G. (2000). *A Molecular Science.* Sudbury, MA: Jones and Bartlett.

Russell, J. W., Kozma, R. B., Jones, T., Wykoff, J., Marx, N., and Davis, J. Use of simultaneous-synchronized macroscopic, microscopic, and symbolic representations to enhance the teaching and learning of chemical concepts. *Journal of Chemical Education.* 74, 330.

Silberberg, M. (1995). *Chemistry.* Orlando, FL: Mosby

Wandersee, J. H., Mintzes, J. J., and Novak, J. D. (1994). Alternative Conceptions in Science. In D. L. Gabel (Ed.) *Handbook of Research on Science Teaching and Learning.* NY: Macmillan.

Williamson, V. M. and Abraham, M. R. (1995). The effects of computer animation on the particulate mental models of college students. *Journal of Research in Science Teaching.* 32, 521.

Part

II

TEACHING STRATEGIES

The Role of Analogies in Chemistry Teaching

MaryKay Orgill
Department of Chemistry
University of Nevada, Las Vegas

George Bodner
Department of Chemistry
Purdue University

Abstract

Many chemistry *teachers* know that they can use analogies to help their students understand challenging or abstract information. Many chemistry *students*, on the other hand, know that analogies can generate a lot of confusion. In this chapter, we will discuss the potential advantages and disadvantages of using analogies in a chemistry classroom. We will also discuss three models that instructors can follow to use analogies effectively in their classes.

Biographies

MaryKay Orgill is an Assistant Professor of Chemistry at the University of Nevada, Las Vegas. Her education has been eclectic and her professional life somewhat schizophrenic so far. She studied chemistry at Brigham Young University to prove that a girl could "do chemistry" and do it well. She was surprised to find during her undergraduate studies that she actually liked chemistry—and loved teaching it. Not willing to be tied to only one kind of learning or working, she enrolled in graduate school at Purdue University to study both biochemistry and chemical education, completing a degree in each of those fields. She continued to pursue both interests as a first-year faculty member with a joint appointment in biochemistry and science education at the University of Missouri-Columbia. During that year, she took on the extra challenge (and incredible learning experience) of teaching a high school chemistry class. In 2004, she moved to her home state of Nevada to take a position at UNLV, where her research focuses on undergraduate chemistry and biochemistry education.

George Bodner is the Arthur E. Kelly Distinguished Professor of Chemistry, Education, and Engineering at Purdue University. He began his academic career as a history/philosophy major at the institution now known as the University at Buffalo. He found, much to his amazement, that chemistry was fun, and he changed his major under the mistaken impression that jobs were easier to find as a chemist. After a mediocre career as an undergraduate (B. S., 1969), he entered graduate school at Indiana University (Ph.D., 1972) where he apparently did well enough as a double major in inorganic and organic chemistry to gain an appointment as a visiting assistant professor at the University of Illinois (1972–1975). Two things became self-evident during his tenure at Illinois: he found that teaching was fun, and he realized that his research could best be described as searching for definitive answers to questions that no one ever asked. When the time came to leave Illinois, he therefore took a job as two-thirds of the chemistry faculty at Stephens College where he lasted for two years (1975–1977), teaching general, organic, inorganic, and biochemistry. He moved to Purdue University in 1977 to take a position in something known as "chemical education." He is the author of more than 100 papers and 45 books or laboratory manuals. His interests include the development of materials to assist undergraduate instruction, research on how students learn, and the history and philosophy of science. Last year, he was

90

selected to receive the Nyholm Medal from the Royal Society of Chemistry, the Pimentel Award in Chemical Education from the American Chemical Society, and the Distinguished Alumni Award from his alma mater, the University at Buffalo.

Introduction

Whether we consciously realize it or not, analogy pervades our existence and our everyday reasoning. We live in a world of "perpetual novelty" (Gentner and Holyoak, 1997). No situation we encounter is exactly like a situation we have encountered previously, and our ability to learn and survive in the world is based on our ability to find similarities between past and present situations and use the knowledge we have gained from past situations to manage current situations. Analogy is powerful in that it allows us to create similarities for a variety of purposes, such as solving problems, creating explanations, or constructing arguments. In particular, an analogy's potential to make explanations of new material intelligible to students by comparing them to material that is already familiar makes analogy a powerful tool for educational purposes.

We've all sat in classes in which a teacher made a difficult or abstract concept understandable by using an analogy. Chemistry classes are full of abstract or challenging concepts that are not easy to understand unless they are related to something from our everyday experiences. For example, as the first author progressed through chemistry courses in her undergraduate studies, she found the concept of hybrid resonance difficult to understand. Her instructors would draw two different representations of benzene on the board, connected by double-headed arrows; this made her think that the resonance hybrid structure was alternating between two different forms. Fortunately, during one class period, an instructor compared a resonance hybrid to a mule. A mule is the product of a cross between a donkey and a horse, and, yet, it has its own unique characteristics. It is neither horse nor donkey, nor does it alternate between being a horse and a donkey. In the same way, a resonance hybrid can be thought of as the product of a cross between two resonance structures. The hybrid is neither of the two contributing structures, but a structure between that of the resonance structures and with its own unique characteristics. This analogy and other effective analogies clarify thinking, help students overcome misconceptions, and give students ways to visualize abstract concepts.

On the other hand, each of us has also sat in a class in which we did not understand an instructor's analogy. In the best-case scenario, we would ignore the analogy; it simply became a waste of class time. In the worst-case scenario, the analogy would confuse, mislead, or keep us from learning class material. While the first author was in graduate school, she was a teaching assistant for a general chemistry class. The instructor of the class had not taught undergraduates previously and, in an attempt to connect with the students on their level, used many analogies during class. Unfortunately, the students (and their teaching assistant!) did not understand these analogies. Their confusion about the analogies kept them from listening to and learning from the other information that was being presented in the class. Their weekly discussion periods were often spent relearning the concepts that they did not understand or could not focus on in the lecture. In this case, the analogies actually hindered learning.

The two chemistry instructors we have mentioned here were well intentioned. They both used analogies to help their students understand abstract and challenging concepts. However, one analogy was more effective than the other in helping students learn. What was the difference between the two? Is there a way to ensure that the analogies you use will be successful? While there is no way to guarantee that the analogies you use in your chemistry class will be understood by all of your students, there are some steps you can take to improve their effectiveness. First, though, you must understand what an analogy is, when analogies are useful in educational settings, and what the potential advantages and disadvantages of using an analogy are.

What Is an Analogy?

Simply put, an analogy is a comparison between two domains of knowledge—one that is familiar and one that is less familiar. The familiar domain is often referred to as the "vehicle," "base," "source," or "analog" domain; the less familiar domain, or the domain to be learned, is usually referred to as the "target" domain. This chapter will use the terms "analog" and "target," respectively, to refer to the two concepts or domains. For example, in many chemistry classes, providing the activation energy needed in order for a reaction to occur is compared to pushing a ball up one side of a hill before letting it roll down the other side. In this example, the person pushing the ball up the hill is the analog concept and activation energy is the target concept. Similarly, in biochemistry textbooks, the enzyme/substrate interaction is compared to placing a key in a lock, where the enzyme/substrate interaction is the target concept and the placement of the key in the lock is the analog concept.

To say that an analogy is a comparison may be an oversimplification. An analogy is not just a comparison between different domains: it is a special kind of comparison that is defined by its purpose and by the type of information it relates. According to Gentner (1989), an analogy is a mapping of knowledge between two domains such that the system of relationships that holds among the objects in the analog domain also holds among the objects in the target domain. Thus, the purpose of an analogy is to transfer a system of relationships from a familiar domain to one that is less familiar (Mason and Sorzio, 1996). The strength of an analogy, therefore, lies less in the number of features the analog and target domains have in common than in the overlap of relational structure between the two domains (Gentner, 1983). For example, the strength of the lock-and-key analogy for enzyme/substrate complementarity is not simply in the fact that the lock corresponds to the enzyme and the key corresponds to the substrate. The strength of that particular analogy is that the relationships between the lock and the key (for example, the shape of the key is complementary to the shape of the lock, and part of the key fits inside the lock) correspond to relationships between the enzyme and the substrate (the shape of the substrate is complementary to the shape of the enzyme, and part of the substrate fits "inside" the enzyme).

Are Analogies Beneficial in Educational Settings?

Very little research has been done about the use of analogies in chemistry classes, and the results of research on whether analogies are beneficial in science education are ambiguous (Beall, 1999). Many studies have reported that using analogies resulted in beneficial outcomes (Beveridge and Parkins, 1987; Brown and Clement, 1989; Cardinale, 1993; Clement, 1993; Donnelly and McDaniel, 1993; Fast, 1999; Glynn and Takahashi, 1998; Harrison and Treagust, 1993; Hayes and Tierney, 1982; Holyoak and Koh, 1987; Simons, 1984; Solomon, 1994; Treagust, Harrison, and Venville, 1996). In a study by Harrison and Treagust (1993), for example, a teacher explained what happens to light when it obliquely enters a more dense medium (refraction) by comparison with what happens to a set of Lego wheels when they roll, unaided, from a hard floor onto a carpeted surface. The trajectory of the light (wheels) is bent toward the normal as it passes through a more dense medium (the carpet) because the light (the wheels) slows down. After the instruction, the students were interviewed, and each seemed, in general, to understand the concepts being taught—both the analogical concept and the target concept in optics. In addition, most of the students were able to transfer their analogical reasoning to a completely new situation. They were able to correctly predict what will happen to light as it moves from a more dense medium to a less dense medium (it bends away from the normal).

Other studies have reported that the use of analogies has had little or no effect on learning (Bean, Searles, and Cowen, 1990; Friedel, Gabel, and Samuel, 1990; Gilbert, 1989). Friedel, Gabel, and Samuel (1990), for example, studied preparatory college chemistry students over the course of two years. Half of the students in their study were instructed with analogies in addition to the regular instruction. Students were given tests to rate their math anxiety, their reasoning abilities, and their visualization abilities.

At the end of the semester, each student took a final exam and a matching exam, in which students were asked to match chemical terms with their analogical corollaries. For example, the following matched questions compare the bags of oranges and moles of Neon:

1. How much would 120 oranges weigh? What would be the mass of 3.60×10^{24} atoms of Ne?
2. What is the weight of 4 bags of oranges? If you had 0.25 moles of Neon, what mass would you have? (Friedel, Gabel, and Samuel, 1990, p. 680).

There were no differences in the posttest scores of the two groups. However, scores on posttests showed that

students in the treatment group who had high visualization skills were actually penalized by using analogs. The data analysis shows that these kinds of students became more successful problem solvers by solving additional practice problems rather than by using analogs. (Friedel, Gabel, and Samuel, 1990, p. 678)

There are two ways to explain why some studies suggest positive results when analogies are used while others show either no effect or a negative effect: analogies are only beneficial under certain circumstances or analogies are only useful for certain kinds of students.

Are Analogies Only Useful in Promoting Learning Under Certain Circumstances? Many reports indicate that analogies may only be useful for teaching target concepts that are conceptually difficult or abstract (Cardinale, 1993; Duit, 1991). If target concepts are relatively simple to understand, an analogy may not be necessary to explain the concepts. In fact, in that case, an analogy may be simply extra information for students to remember (Gick and Holyoak, 1983). In chemistry, however, where concepts are often novel and challenging or difficult to visualize, the use of analogies may have beneficial effects on learning (Harrison and Treagust, 1996).

There is also reason to believe that certain instructional criteria must be met in order for an analogy to be effective. Gabel and Samuel (1986), for example, found that analogies are most useful when students understand the analog domain well. In the case of Gabel and Samuel's article, the concentration of lemonade was compared to the concentration of other chemical solutions. Students were able to use the lemonade analogy to solve problems in which the concentration of a solution was changed by adding solvent because they had experienced diluting a strong lemonade drink by adding water. On the other hand, students did not find the lemonade analogy as useful when solving problems in which the concentration of a solution was changed by evaporating solvent because the students were not familiar with making a weak lemonade drink stronger by evaporating off some of the water.

Other researchers have suggested that effective use of analogies occurs when teachers explicitly compare analog and target domains and identify the limitations of an analogy (see Glynn, 1991; Treagust, 1993; Zeitoun, 1984). Teachers can even guide their students through the identification of these similarities and limitations. We have used this strategy in a high school chemistry class. The students were beginning a unit about chemical reactions and chemical equations, and we used an analogy that compared chemical equations to recipes to introduce the concepts. We gave the students a page on which we printed both a recipe and a chemical reaction and asked the students to identify the ways in which equations are similar to chemical recipes. They were able to identify several similarities, which each student noted on his or her own page; some of these similarities are as follows: (1) some equations have names and some recipes have names; (2) equations list reactants (chemicals that will be added together) and recipes list ingedients (foods that will be added together); (3) equations list the physical state of the reactants and recipes list the physical state of ingredients (mashed versus sliced, for example); and (4) equations list the conditions under which a reaction takes place and recipes list the conditions under which baking or cooking occurs. After we discussed the similarities between equations and recipes, we asked the students to identify the ways in which equations and recipes are different. They were able to determine that while recipes give the time needed for a recipe to be completed, chemical equations do not indicate how long a chemical reaction will take to occur. Because the students understood and wrote down the meaning of the analogy, they refered to the analogy during subsequent learning about chemical equations and reactions.

What Kind of Students Benefit from Analogies? The research literature suggests that not all analogies are useful analogies. Even a "good" analogy may not be useful for all students. Several studies suggest that analogies are more useful for low-ability students than for high-ability students (Bean, Singer, and Cowan, 1985; Donnelly and McDaniel, 1993; Duit, 1991; Gabel and Sherwood, 1980). Studies by Gabel and Sherwood (1980) and Donnelly and McDaniel (1993) indicated that instruction in analogies seemed to be more helpful for students of low formal reasoning ability and high mathematics anxiety than for more capable students because the achievement scores of students with lower formal reasoning ability changed more after analogy instruction than the achievement scores of students with higher ability.

The work which suggests that analogies are more useful for low-ability students could be deceiving, however. It is possible that no change was seen in the achievement scores of the high-ability students because their scores were closer to the maximum available score before instruction with analogies. It is also possible that the high-ability students had a good understanding of the material before analogy instruction, in which case instruction with analogies would not significantly improve their understanding of the concept.

Regardless of whether low-ability students are, in fact, helped more by analogies than high-ability students, there is evidence that some teachers believe this is the case and tend to use more analogies with students they consider to be of lower reasoning abilities. When student teachers were interviewed about their use of analogies, they indicated that they tended to use more analogies with students they perceived as having lower reasoning abilities than with students they perceived as having higher reasoning abilities (Jarman, 1996).

Whether an analogy is useful to a given student may also depend on the student's familiarity with the topic being taught. Novick (1988) divided undergraduate students into groups of "experts" and "novices" according to their math SAT scores and gave them a target problem to solve:

> Members of the West High School Band were hard at work practicing for the annual Homecoming Parade. First they tried marching in rows of twelve, but Andrew was left by himself to bring up the rear. The band director was annoyed because it didn't look good to have one row with only a single person in it, and of course Andrew wasn't very pleased either. To get rid of this problem, the director told the band members to march in columns of eight. But Andrew was still left to march alone. Even when the band marched in rows of three, Andrew was left out. Finally, in exasperation, Andrew told the band director that they should march in rows of five in order to have all the rows filled. He was right. This time all the rows were filled and Andrew wasn't alone any more. Given that there were at least 45 musicians on the field but fewer than 200 musicians, how many students were there in the West High school Band? (Novick, 1988, p. 513)

Half of the experts and half of the novices were also given an analogous problem that potentially could help them solve the target problem:

> Mr. and Mrs. Renshaw were planning how to arrange vegetable plants in their new garden. They agreed on the total number of plants to buy, but not on how many of each kind to get. Mr. Renshaw wanted to have a few kinds of vegetables and ten of each kind. Mrs. Renshaw wanted more different kinds of vegetables, so she suggested having only four of each kind. Mr. Renshaw didn't like that because if some of the plants died, there wouldn't be very many left of each kind. So they agreed to have five of each vegetable. But then their daughter pointed out that there was room in the garden for two more plants, although then there wouldn't be the same number of each kind of vegetable. To remedy this, she suggested buying six of each vegetable. Everyone was satisfied with this plan. Given this information, what is the fewest number of vegetable plants the Renshaws could have in their garden? (Novick, 1988, p. 513)

The subjects were given a method for solving the analogous problem, but not a method for solving the target problem. The novices were not affected by seeing the analogous problem, but the experts demonstrated positive transfer from the analogous problem to the target problem.

In another situation, the first author observed a biochemistry class in which an instructor shared an analogy comparing the process of putting a hand in a rubber glove to the induced fit model of enzyme/substrate binding. The author thought the analogy was a wonderful way to visualize the flexibility of enzymes and their ability to adapt their shapes to those of their substrates (just as the shape of the glove adapted to the shape of the hand). She discovered, however, that the analogy was not as useful to the students as it was to her. The students understood that the point of the analogy was to convey the complementarity of the shapes of enzymes and substrates. The students, with their limited understanding of biochemistry, were not able to recognize the purpose of the analogy or use the analogy to the degree intended by their instructor.

Conflicting results were seen in a study by Donnelly and McDaniel (1993), in which students who were learning a previously unfamiliar scientific concept ("novices" in the terms of the Novick study) were divided into two groups. One group was taught with analogies; the other was not. The novice students who were taught with analogies outperformed their peers who were not.

The apparent inconsistency between the results obtained by Novick and by Donnelly and McDaniel can be explained by arguing that experts may be able to recognize and use analogies more easily than novices, but novices may benefit more from the use of analogies than experts. The challenge for teachers is to determine how to help novices recognize and use analogies for their benefit. After all, even students with high-level reasoning abilities can be novices in a field in which information is new to them.

Potential Beneficial Roles of Analogy

Analogies are most often used in an educational setting to help students understand new information in terms of already familiar information and to help them relate that new information to their already existing knowledge structure (Beall, 1999; Glynn, 1991; Simons, 1984; Thiele and Treagust, 1991; Venville and Treagust, 1997). It has been argued that "knowledge is constructed in the mind of the learner" (Bodner, 1986, p. 873). As they construct knowledge, learners seek to give meaning to the information they are learning, and the comparative nature of analogies promotes such meaningful learning. Ausubel, Novak, and Hanesian (1978) state that in order to learn meaningfully, individuals must choose to relate new knowledge to concepts they already know.

By their very nature, analogies relate information in a familiar, analog domain to information in an unfamiliar, target domain. Lemke notes:

> What makes an analogy work is very simple in thematic terms. An analogy sets up a simple correspondence between two thematic patterns. The patterns have different thematic items, but the same semantic relations between them. One pattern is already familiar, the other new. Students learn to transfer semantic relationships from the familiar thematic items and their pattern to the unfamiliar items and their pattern. (Lemke, 1990, p. 117)

There are several roles that analogies can play in promoting meaningful learning. First, they help learners organize information or view information from a new perspective. Thiele and Treagust (1991) argue that analogies help to arrange existing memory and prepare it for new information. Consider an analogy that has been used to help high school chemistry students understand the general organization of the periodic table. The analogy compares the periodic table to the geography of the United States. The United States is divided into different regions—the West, the Midwest, and the East—on the basis of similarities in the geography and weather patterns of the states in each region. Similarly, elements are grouped in the periodic table—as metals, metalloids, and non-metals—based on similarities in their physical and chemical characteristics. Certainly, the analogy could be extended to explain the organization of the periodic table in more detail; but, in the case of beginning high school students, the analogy provides a way for students to mentally organize the information they will learn about the periodic table.

Analogies can also give structure to information being learned by drawing attention to significant features of the target domain (Simons, 1984) or to particular differences between the analog and target domains (Gentner and Markman, 1997). Gick and Holyoak (1983) argue that analogies can "[...] make the novel seem familiar by relating it to prior knowledge [and] make the familiar seem strange by viewing it from a new perspective"(p. 2).

For example, Stephanie, a biochemistry student, had heard in previous classes that DNA is like a blueprint. She had a partial understanding of the analogy, that DNA contains the information needed to create an organism. However, she had never considered the other implications of the analogy, namely that a blueprint is a two-dimensional overhead view of an object to be made. DNA, on the other hand, is not a two-dimensional picture of what is going to be made. Her instructor explained that, in his opinion, DNA is more like a recipe than a blueprint because DNA contains the information needed for making something instead of a picture of something to be made. His explanation caused Stephanie to think about DNA in a different way than she had previously, even though she had a hard time putting her new understanding into words. If nothing else, the new analogy caused Stephanie to look more deeply at her own understanding of the concept:

> Stephanie: I've always been taught in class…they always say blueprint, so [Dr. Carter's analogy] actually opened my eyes that it's really not like a blueprint because I always think…I don't know…my definition of blueprint, I kind of think, "OK…here's what has to be made and, like, the cell's going to use," like a blue…it's kind of like a blueprint. I know it's not exactly like a blueprint, but you can see…if you think of the word "blueprint," you're like, "OK. Here's what has to be made and then the cell reads it and then it makes it." So, it's like…I guess I kind of think of a blueprint of how much detail a blueprint actually goes into and, then, yeah, it would be more like a recipe if you actually

thought about how in depth a blueprint really is. I was like, "a blueprint is the copy and then you figure out what it is from that." So, at least it changed my thinking that it's not really like a blueprint.

Analogies may also help students visualize abstract concepts, orders of magnitude, or unobservable phenomena (Dagher 1995a; Harrison and Treagust, 1993; Simons, 1984; Thiele and Treagust, 1994; Venville and Treagust, 1997). When they do this, they can provide a concrete reference that students can use when thinking about challenging, abstract information (Brown, 1993; Simons, 1984). One of the difficult concepts for a beginning chemistry student to understand is the relative size of an atom and its nucleus. A teacher's saying that the atom is about 100,000 times larger than the nucleus may not have any physical reality for a student. However, comparing the relative size of a nucleus and an atom to the relative size of a marble in a football stadium may give the student a way to visualize the concept.

Analogies can play a motivational role in meaningful learning (Bean, Searles, and Cowen, 1990; Dagher 1995a; Glynn and Takahashi, 1998; Thiele and Treagust, 1994). The use of analogies can result in better student engagement and interaction with a topic. Lemke (1990) asserts that students are three to four times more likely to pay attention to the familiar language of an analogy than to unfamiliar scientific language. The familiar language of an analogy can also give students who are unfamiliar or uncomfortable with scientific terms a way to express their understanding of and interact with a target concept. Dagher (1995a) argues that the language of analogies can demystify scientific language and notes that the use of narrative analogies tends to result in higher student motivation and engagement.

Motivation is not only a product of the students' interest in a topic, but also of their beliefs about their abilities to successfully understand or solve a problem in that topic area; and analogies can affect both of these contributors to motivation. Analogies can make new material interesting to students, particularly when the analogy relates new information to the students' real-world experiences (Thiele and Treagust, 1994). They can also increase students' beliefs about their problem-solving abilities. Although students may initially believe themselves incapable of solving a new problem or of understanding new information, their beliefs about their abilities may change when the new problem or new information is related by analogy to a problem or information they have already been successful in solving or understanding (Pintrich, Marx, and Boyle, 1993).

Students we have interviewed mention that when instructors use analogies in class, they are indicating their concern for their students and their learning. Likewise, the instructors we have interviewed believe that good instructors make information understandable through analogies. The students' perception of their instructors' concern for them seems to motivate them to study and learn.

Finally, as mentioned earlier in this chapter, analogies can play a role in promoting conceptual change by helping students overcome existing misconceptions (Brown, 1992, 1993; Brown and Clement, 1989; Clement, 1993; Dagher, 1994; Dupin and Johsua, 1989; Gentner et al., 1997; Mason, 1994; Venville and Treagust, 1996). Ideally, analogies can help students recognize errors in conceptions they currently hold, reject those conceptions, and adopt new conceptions that are in line with those accepted by the scientific community. Analogies may make new ideas intelligible and initially plausible by relating them to already familiar information. If students can assimilate new information in terms of their existing knowledge, they are likely to be able to understand that information, relate it in their own words, and comprehend how that new information might be consistent with reality—all necessary conditions for conceptual change (Posner, Strike, Hewson, and Gertzog, 1982). Conceptual understanding is discussed in Chapter 7 of this book.

Multiple analogies can also play roles in conceptual change. Brown and Clement (1989) have developed the "Bridging Analogies Strategy" to help students overcome misconceptions. In this strategy, instructors first try to make a misconception explicit by asking a target question. They then present a case that they see as analogous and try to establish the similarity/analogy relation. If students do not see how the analogous situation applies to the target situation and do not transfer knowledge from the analogous situation to the target situation, instructors introduce another analogy, one that is conceptually midway between the first analogy and the target concept. This process continues incrementally until the students can see the similarity between the first analogy and the target concept and transfer knowledge from the first analogous situation to the target situation.

In one specific example, Brown and Clement (1989) were trying to help students understand the upward force that a table exerts on a book (which most students could not fathom). They did this by going through a series of analogies ("bridging analogies"), each of which more closely approached the target concept than the previous. They started by asking students if there was an upward force when books rest on an outstretched hand. Although the interviewed student agreed that his hands did exert an upward force on the books, he could not see how this situation was analogous to that of a book sitting on a table. The interviewer introduced another analogy—that of a book resting on a spring. This time, the student did not understand that the spring exerted an upward force, so the interviewer introduced another bridging analogy—that of a hand pushing down on a spring. The student did believe that the spring would exert an upward force on his hand because he saw his hand as actively pushing on the spring while the book resting on the spring was passive.

As an attempt to help the student understand that a spring does, indeed, exert an upward force on a resting book, the interviewer introduced yet another bridging analogy—that of a hand resting on the book which was resting on a spring. This analogy helped the student see that an upward force is exerted by a spring on any object resting on it. However, the goal was to help the student see that a table exerts an upward force on a book. To this end, the interviewer introduced two ideas: a pile of books resting on a flexible board and a hand resting on a flexible board. Although, initially, the student did not think that the flexible table exerted an upward force, he reasoned that the flexible table was similar to the spring and would exert an upward force. Ultimately, he could make the connection between the flexible table exerting an upward force on a book and a table exerting a force on a book.

The series of bridging analogies helped the student to incrementally change his views towards the views of physicists. While this approach was effective for the student, there is also evidence that "experts" use bridging analogies when problem solving to increase their confidence in a problem solution or the problem-solving process (Clement, 1993).

Ideally, if students view the analog and target as analogous and they understand the analog concept, they will change their conception of the target concept. However, this is not always the case. In one case cited by Brown and Clement (1989), those conditions were met, but the student did not change his ideas about the target concept. Although he found the analogous situation intelligible, he did not find it plausible (reflecting the real world), so he did not transfer analog concepts to the target concept. In cases where conceptual change resulted from the use of bridging analogies, the authors note that the analogies helped enrich students' conceptions of the target concepts, and they suggest that this enrichment was necessary to affect conceptual restructuring.

Potential Negative Results of Analogy Use

As with any other teaching technique, the use of analogies in a classroom can have a negative effect. Some of these negative effects can be avoided if teachers follow certain guidelines when teaching with analogies (see Glynn, 1991; Treagust, 1993; Zeitoun, 1984), but at least some of these negative effects are possible even when teachers follow those guidelines. Although both teacher and student may consider an analogy useful for learning new information, the analogy might be superfluous information if the student already has an understanding of the target concept being taught (Venville and Treagust, 1997). In one biochemistry class, an instructor compared hydrogen bonds to Velcro. Individually, hydrogen bonds are weak, but large numbers of hydrogen bonds can act together to stabilize and strengthen a structure. Similarly, although one hook and eye of Velcro is not strong, thousands of Velcro hooks and eyes working together have the strength to hold two materials together. While this analogy might have been useful to someone who is beginning to learn about hydrogen bonds, it was not useful for the students in the biochemistry class who already had a good understanding of the concept of hydrogen bonds.

Students may resort to using the analogy mechanically, without considering the information the analogy was meant to convey (Arber, 1964; Gentner and Gentner, 1983; Venville and Treagust, 1997). For example, a student may answer an exam question with an analogy (Question: "What is the function of the mitochondrion?" Answer: "The mitochondrion is the power plant of the cell."). Part of the mechanical use of analogy may be due to the students' not being willing to invest time to *learn* a concept if they can simply remember a familiar analogy for that concept, since familiar analogies can often provide students with correct answers to exam questions—even if those analogies are not understood (Treagust, Harrison, and Venville, 1996).

A chemistry instructor noted that each year students had difficulty predicting the relative pH of ionic salts when they dissolve in water. One year, he decided to begin a lesson on the hydrolysis of salts by asking students to think of the relationship between dominant and recessive genes in parents in relationship to children. He asked if one parent had brown eyes and the other blue eyes, what eye color would their child most likely have? He next told students that the parents of the salt (an acid and a base) could be classified as being either strong or weak. The child of the parents, the salt, would have pH characteristics of the dominant parent. For example, sodium acetate can be made by reacting sodium hydroxide, a strong base, with acetic acid, a weak acid. When placed in water, the salt solution should be basic because it has a strong base for one of its parents. The students immediately understood this analogy and the class was very successful at predicting the relative pH of salt solutions. The instructor admonished the students that they could not use this analogy when explaining why the solution was acidic, basic, or neutral. However, when asked to explain why certain solutions were acidic, basic, or neutral on an examination, the majority of students cited the strong versus weak parent as a reason.

The mechanical use of an analogy may also be due to students' inability to differentiate the analogy from reality. An analogy never completely describes a target concept. Each analogy has limitations. Unfortunately, students usually do not know enough about the target concept to understand those limitations. For this reason, they may either accept the analogical explanation as a statement of reality about the target concept or incorrectly apply the analogy by taking the analogy too far. Beall, using the word "metaphor" to mean either "metaphor" or "analogy," says that this is often the case in biochemistry and gives a particular example:

> Concepts in biochemistry are very commonly understood using language as a metaphor. For example, a letter is the metaphor for a single amino acid residue in a protein; a word corresponds to the secondary protein structure; and so on, up to a complete book, which corresponds to the entire cell. This metaphor is so attractive that it colors thinking about these subjects and if carried too far can lead to erroneous impressions. (Beall, 1999, p. 367)

When students inappropriately apply irrelevant concepts from the analog domain to the target domain, they can develop misconceptions about the target domain (Brown and Clement, 1989; Clement, 1993; Duit, 1991; Glynn, 1995; Kaufman, Patel, and Magder, 1996; Thagard, 1992; Zook, 1991; Zook and DiVesta, 1991; Zook and Maier, 1994). An analogy that is often used in biochemistry compares a cell to a factory and the different organelles to parts of the factory. Students who know a lot about factories but little about the cell might assume that the cell, like the factory, has a limited number of entrances. These misconceptions that are developed as the result of an analogy can be difficult to remedy.

Finally, although one of the purposes of an analogy is to help students learn a concept meaningfully by relating that concept to the students' prior knowledge, the use of an analogy may limit a student's ability to develop a deep understanding of that concept (Brown, 1989; Dagher, 1995b; Spiro, Feltovich, Coulson, and Anderson, 1989). When only one analogy is used to convey information about a particular topic, students may accept their teacher's analogical explanation as the only possible or necessary explanation for a given topic.

Spiro, Feltovich, Coulson, and Anderson (1989) found that medical students were kept from a full understanding of concepts associated with myocardial failure because of analogies they had learned. They noted:

> [...], although simple analogies rarely if ever form the basis for a full understanding of a newly encountered concept, there is nevertheless a powerful tendency for learners to continue to limit their understanding to just those aspects of the new concept covered by its mapping from the old one. Analogies seduce learners into reducing complex concepts to a simpler and more familiar analogical core. (Spiro, Feltovich, Coulson, and Anderson, 1989, p. 498)

It may simply be more convenient for students to think of a concept as being explained by one familiar analogy than to invest the time to learn a new explanation for or develop a correct understanding of that concept.

Teaching Models

Although analogies can form conceptual bridges between knowledge that students have and new information, their incorrect use can lead the students to develop incorrect ideas about target concepts. Observational studies have shown that teachers often use analogies spontaneously and, usually, unsystematically (Glynn, Duit, and Thiele, 1995; Thiele and Treagust, 1994). Several authors have suggested that teachers could use analogies more effectively if they had guidelines for teaching with analogies. Three major teaching models are presented in the analogy literature: the Teaching-With-Analogies (TWA) model, the General Model of Analogy Teaching (GMAT), and the FAR (Focus, Action, Reflection) model.

Teaching-With-Analogies Model (TWA). The teaching model cited most frequently in the literature is the Teaching-With-Analogies model (Glynn, 1991, 1995, 1996). Glynn developed his guidelines for teaching with analogies by examining what he considered to be exceptional analogies from science textbooks. The Teaching-With-Analogies model outlines six steps that teachers should follow when using analogies as teaching tools. Each step is consistent with factors that have been reported as having positive effects on correct analogical transfer:

- Introduce the target concept,
- Present the analog concept (a concept with which the students should be familiar from previous experience),
- Identify the relevant features of the target and analog concepts,
- Explicitly map the similarities between the target and analog concepts,
- Indicate where the analogy breaks down, and
- Draw conclusions about the target concept based on the analog concept.

While these steps do not need to be followed in any certain order, teachers should include the features of each of the six steps outlined above in any discussions that include analogies.

Although the TWA model is mentioned extensively in the analogy literature, relatively few studies have examined its effectiveness. Treagust, Harrison, and Venville (1996) tutored seven high school teachers in the TWA model of analogy instruction and then observed sessions in which the teachers used analogies and comparable teaching sessions in which the same concepts were taught without the use of analogies. After the teaching sessions, they interviewed students and teachers about the concepts that were taught and examined interview transcripts for evidence of conceptual change. In particular, they looked for statements that would indicate that the students found their explanations for certain phenomena as intelligible, plausible, or fruitful (Posner, Strike, Hewson, and Gertzog, 1982; Strike and Posner, 1985).
Treagust, Harrison, and Venville (1996) determined that students who were taught with an analogy by the TWA model demonstrated a higher-level conception status than students who were not taught with the analogy. In each of the three case studies described, the authors felt that the use of the analogy was an essential link to the students' ability to make sense out of phenomena. Only one of the students in the case studies used the analogy spontaneously, but the other two students made "conceptual progress" when reminded about the analogy. It appears that analogies can, indeed, promote meaningful learning and conceptual growth when used systematically and in accordance with the TWA model.

General Model of Analogy Teaching (GMAT). Zeitoun's General Model of Analogy Teaching (GMAT) differs from the TWA model in that it describes additional pedagogical aspects of teaching with analogies (Zeitoun, 1984). Zeitoun's model emphasizes the need to plan analogies before using them, to take into account students' prior knowledge and abilities, to evaluate the effects of the analogy, and to revise the analogy to meet the needs of the students. The GMAT model consists of the following steps:

- Measure some of the students' characteristics related to analogical learning in general;
- Assess the prior knowledge of the students about the topic;
- Analyze the learning material of the topic;
- Judge the appropriateness of the analogy to be used;
- Determine the characteristics of the analogy to be used;

- Select the strategy of teaching and the medium of presenting the analogy;
- Present the analogy to the students (including its purpose, the analogous attributes, the transfer statements, and the irrelevant attributes);
- Evaluate the outcomes of using the analogy in teaching (determine whether students use the analogy to study the topic, assess the students' knowledge of the attributes of the topic, and identify the misconceptions that result from the analogy); and
- Revise the stages of the model if needed.

Zeitoun claims that analogies will be used more effectively and with less misconceptions if teachers follow his guidelines, but we have not seen any reports of studies of the effectiveness of this model.

FAR (Focus, Action, Reflection) Model. Treagust and his colleagues (Treagust, 1993; Treagust, Harrison, and Venville, 1998) developed their FAR (Focus, Action, Reflection) model after observing five experienced teachers who used the TWA model with their favorite analogies. They found that although these experienced teachers did use each of the steps of the TWA model of teaching with analogies when they taught, they did not use the steps in any consistent order. Instead, they modified the order of the steps to meet the needs of their students and of the lesson they were teaching. These teachers also spent some time preparing their analogies before instruction and reflecting on the effects of using the analogy after instruction—actions that Treagust, Harrison, and Venville felt were necessary for the teachers' effective use of analogies. Accordingly, the FAR guide integrates preparation and reflection stages into the actual instruction stage of using analogies.

The FAR guide is simpler than either the TWA or GMAT models and is so by design. The developers of the FAR guide felt that there were too many steps to remember in the TWA and GMAT models, so they wanted to develop a guide for teaching with analogies that any teacher could remember easily (Treagust, 1993; Treagust, Harrison, and Venville, 1998). The steps of their FAR guide are found below (Treagust, 1993, p. 299):

FOCUS. on the concept being taught and the analog to be used. Is it difficult, unfamiliar, or abstract? What do students know about the concept? Are students familiar with the analog?

ACTION. Explicitly connect the similarities between the analog and target concepts and discuss the limitations of the analogy.

REFLECTION. Evaluate how the analogy came across to the students and make improvements as needed.

The effects of using the FAR guide have not been investigated; however, there is one example in which the FAR guide was successfully used to teach a topic with analogies. Harrison and Treagust (2000) observed 11th-grade chemistry students who were taught about atoms and molecules by their regular classroom teachers who used analogies in a systematic way, with reference to the FAR guide. All formal and informal discussions about the topics were taped, and the investigators collected student work and interviewed students in order to determine their conceptions. The authors present a case study of one of the students, who they call "Alex, the multiple modeler." The way that Alex used the multiple analogies/models of atomic structure throughout the class provided evidence that he had changed his initial conceptions about atomic structure in favor of more scientific conceptions. Initially, Alex believed that an atom was composed of a large nucleus with closely-situated, orbiting electrons. However, after instruction with analogical models, Alex used multiple models to describe his new conception of an atom as consisting of a central nucleus surrounded by spacious (more spacious than his original description), swirling electron clouds.

Summary

There are several potential advantages to using analogies in a chemistry classroom. Analogies can help students visualize abstract concepts, organize their thinking about a given topic, and learn a topic meaningfully. They can also motivate students to learn. There is always, however, a danger that analogies will be misinterpreted or misunderstood by students. Teachers often use analogies unsystematically in their classroom teaching, and that unsystematic use may result in the development of misconceptions about target concepts or, at the least, less effective analogical transfer than is possible. Several authors have suggested models by which analogies can be taught effectively. Following these models and understanding the advantages and disadvantages of analogy use may help instructors to use them more effectively in the chemistry classroom.

Effectively used analogies can help students understand difficult concepts, often with surprising results. Earlier this year, we explained the concepts of compounds, elements, and mixtures to our high school students. We defined the different systems and drew pictures on the chalkboard representing microscopic views of these systems; however, the students did not seem to understand the differences between them. In an attempt to explain the systems, we told the students about a cereal analogy for compounds, elements, and mixtures.

In this analogy, mixtures are compared to Raisin Bran cereal because it contains two separate components (the raisins and the flakes), and the composition of a sample of Raisin Bran differs depending on where you take the sample: if you take the sample at the bottom of the box, you will get more raisins than if you take a sample from the top of the box. Compounds are compared to Crispix cereal because each time you reach into a Crispix box, you will pull out the same pieces: a "bonded" square made of rice on one side and corn on the other.

Having described the analogies for compounds and mixtures, we asked the students to identify cereals that would be analogies for elements. The students' response was incredible. Students who do not normally participate in class discussion were volunteering cereals that could be called analogical "elements": cheerios, fruit loops (if you ignore the colors), and corn flakes. One student, who struggles in chemistry class, raised his hand and used an analogy to check his understanding of the definition of "compound." He asked, "Well, would Frosted Mini-Wheats be a compound?" When we asked him to explain what he meant, he said, "When you reach into the box, you always get the same things, but each thing in the box is made up of two parts: a frosted side and an unfrosted side." For this particular student, this conclusion was brilliant. Although he did not understand the initial description of compounds, the analogy helped him make sense of what, for him, was a difficult concept. You can see similar results from using analogies in your chemistry class when you plan your analogies and carefully explain them in class. The students like analogies, and they do use well-explained analogies to learn!

Suggested Readings

Glynn, S. M. (1991). Explaining science concepts: A teaching with analogies model. In S. Glynn, R. Yeany, and B. Britton (Eds.) *The psychology of learning science*. 219-240. Hillsdale, NJ: Erlbaum.

Glynn, S. (1995). Conceptual bridges: Using analogies to explain scientific concepts. *Science Teacher*. 62, 24-27.

Clement, J. (1993). Using bridging analogies and anchoring intuitions to deal with students' preconceptions in physics. *Journal of Research in Science Teaching*. 30, 1241-1257.

References

Arber, A. (1964). *The mind and the eye: A study of the biologist's standpoint*. Cambridge, MA: Cambridge University Press.

Ausubel, D. P., Novak, J. D., and Hanesian, H. (1978). *Educational psychology: A cognitive view*. London: Holt, Rinehart, and Winston.

Beall, H. (1999). The ubiquitous metaphors of chemistry teaching. *Journal of Chemical Education*. 76, 366–368.

Bean, T. W., Searles, D., and Cowen, S. (1990). Text-based analogies. *Reading Psychology*. 11, 323–333.

Bean, T. W., Singer, H., and Cowen, S. (1985). Analogical study guides: Improving comprehension in science. *Journal of Reading*. 29, 246–250.

Beveridge, M., and Parkins, E. (1987). Visual representation in analogical problem solving. *Memory and Cognition.* 15, 230–237.

Bodner, G. M. (1986). Constructivism: A theory of knowledge. *Journal of Chemical Education.* 63, 873 – 877.

Brown, A. (1989). Analogical learning and transfer: What develops? In S. Vosniadou, and A. Ortony (Eds.). *Similarity and analogical reasoning.* 369–412. Cambridge, MA: Cambridge University Press.

Brown, D. E. (1992), Using examples and analogies to remediate misconceptions in physics: Factors influencing conceptual change. *Journal of Research in Science Teaching.* 29, 17–34.

Brown, D. E. (1993), Refocusing core intuitions: A concretizing role for analogy in conceptual change. *Journal of Research in Science Teaching.* 30, 1273–1290.

Brown, D., and Clement, J. (1989). Overcoming misconceptions via analogical reasoning: Abstract transfer versus explanatory model construction. *Instructional Science.* 18, 237–261.

Cardinale, L. A. (1993). Facilitating science by learning by embedded explication. *Instructional Science.* 21, 501–512.

Clement, J. (1993). Using bridging analogies and anchoring intuitions to deal with students' preconceptions in physics. *Journal of Research in Science Teaching.* 30, 1241–1257.

Dagher, Z. R. (1994). Does the use of analogies contribute to conceptual change? *Science Education.* 78, 601–614.

Dagher, Z. R. (1995a) Analysis of analogies used by science teachers. *Journal of Research in Science Teaching.* 32, 259–270.

Dagher, Z. R. (1995b). Review of studies on the effectiveness of instructional analogies in science education. *Science Education.* 79, 295–312.

Donnelly, C. M., and McDaniel, M. A. (1993). Use of analogy in learning scientific concepts. *Journal of Experimental Psychology: Learning, Memory, and Cognition.* 19, 975–987.

Duit, R. (1991). On the role of analogies and metaphors in learning science. *Science Education.* 75, 649–672.

Dupin, J., and Johsua, S. (1989). Analogies and "modeling analogies" in teaching some examples in basic electricity. *Science Education.* 73, 207–224.

Fast, G. R. (1999). Analogies and reconstruction of probability knowledge. *School Science and Mathematics.* 99, 230–240.

Friedel, A. W., Gabel, D. L., and Samuel, J. (1990). Using analogs for chemistry solving: Does it increase understanding? *School Science and Mathematics.* 90, 674–682.

Gabel, D. L., and Samuel, K. V. (1986). High school students' ability to solve molarity problems and their analog counterparts. *Journal of Research in Science Teaching.* 23, 165–176.

Gabel, D. L., and Sherwood, R. D. (1980). Effect of using analogies on chemistry achievement according to Piagetian level. *Science Education.* 64, 709–716.

Gentner, D. (1983). Structure-mapping: A theoretical framework for analogy. *Cognitive Science.* 7, 155–170.

Gentner, D. (1989). The mechanisms of analogical learning. In S. Vosniadou and A. Ortony (Eds.) *Similarity and analogical reasoning.* 199–241. Cambridge, MA: Cambridge University Press.

Gentner, D., Brem, S., Ferguson, R. W., Markman, A. B., Levidow, B. B., Wolff, P., and Forbus, K. D. (1997). Analogical reasoning and conceptual change: A case study of Johannes Kepler. *Journal of the Learning Sciences.* 6, 3–40.

Gentner, D., and Gentner, D. R. (1983). Flowing waters or teeming crowds: Mental models of electricity. In D. Gentner and A. L. Stevens (Eds.) *Mental models.* 99-129. Hillsdale, NJ: Lawrence Erlbaum.

Gentner, D., and Holyoak, K. J. (1997). Reasoning and learning by analogy—Introduction. *American Psychologist.* 52, 32–34.

Gentner, D., and Markman, A. B. (1997). Structure mapping in analogy and similarity. *American Psychologist.* 52, 45–56.

Gick, M. L., and Holyoak, K. J. (1983). Schema induction and analogical transfer. *Cognitive Psychology.* 15, 1–38.

Gilbert, S. W. (1989). An evaluation of the use of analogy, simile, and metaphor in science texts. *Journal of Research in Science Teaching.* 26, 315–327.

Glynn, S. M. (1991). Explaining science concepts: A teaching with analogies model. In S. Glynn, R. Yeany, and B. Britton (Eds.) *The psychology of learning science.* 219–240. Hillsdale, NJ: Erlbaum.

Glynn, S. (1995). Conceptual bridges: Using analogies to explain scientific concepts. *Science Teacher.* 62, 24–27.

Glynn, S. (1996). Teaching with analogies: Building on the science textbook. National Reading Research Center. *Reading Teacher.* 49, 490–492.

Glynn, S. M., Duit, R., and Thiele, R. B. (1995). Teaching science with analogies: A strategy for constructing knowledge. In S. M. Glynn and R. Duit (Eds.) *Learning science in the schools: Research reforming practice.* 247–276. Mahwah, NJ: Erlbaum.

Glynn, S. M. and Takahashi, T. (1998). Learning from analogy-enhanced science text. *Journal of Research in Science Teaching.* 35, 1129–1149.

Harrison, A. G.. and Treagust, D. F. (1993). Teaching with analogies: A case study in grade-10 optics. *Journal of Research in Science Teaching.* 30, 1291–1307.

Harrison, A. G.. and Treagust, D. F. (1996). Secondary students' mental models of atoms and molecules: Implications for teaching chemistry. *Science Education.* 80, 509–534.

Harrison, A. G., and Treagust, D. F. (2000). Learning about atoms, molecules, and chemical bonds: A case study of multiple-model use in grade-11 chemistry. *Science Education.* 84, 352–381.

Hayes, D. A., and Tierney, R. J. (1982). Developing readers' knowledge through analogy. *Reading Research Quarterly.* 17, 256–280.

Holyoak, K. J., and Koh, K. (1987). Surface and structural similarity in analogical transfer. *Memory and Cognition.* 15, 332–340.

Jarman, R. (1996). Student teachers' use of analogies in science instruction. *International Journal of Science Education.* 18, 869–880.

Kaufman, D. R., Patel, V. L., and Magder, S. A. (1996). The explanatory role of spontaneously generated analogies in reasoning about physiological concepts. *International Journal of Science Education.* 18, 369–386.

Lemke, J. L. (1990). *Talking science: Language, learning, and values.* Norwood, NJ: Ablex Publishing Corp.

Mason, L. (1994). Cognitive and metacognitive aspects in conceptual change by analogy. *Instructional Science.* 22, 157–187.

Mason, L., and Sorzio, P. (1996). Analogical reasoning in restructuring scientific knowledge. *European Journal of Psychology of Education.* 11, 3–23.

Novick, L. R. (1988). Analogical transfer, problem similarity, and expertise. *Journal of Experimental Psychology: Learning, Memory, and Cognition.* i, 510–520.

Pintrich, P. R., Marx, R. W.. and Boyle, R. A. (1993). Beyond cold conceptual change: The role of motivational beliefs and classroom contextual factors in the process of conceptual change. *Review of Educational Research.* 63, 167–199.

Posner, G. J., Strike, K. A., Hewson, P. W., and Gertzog, W. A. (1982). Accommodation of scientific conception: Toward a theory of conceptual change. *Science Education.* 66, 211–227.

Simons, P. R. J. (1984). Instructing with analogies. *Journal of Educational Psychology.* 76, 513–527.

Solomon, I. (1994). Analogical transfer and functional fixedness in the science classroom. *Journal of Educational Research.* 87, 371–377.

Spiro, R. J., Feltovich, P. J., Coulson, R. L., and Anderson, D. K. (1989). Multiple analogies for complex concepts: antidotes for analogy-induced misconception in advanced knowledge acquisition. In S. Vosniadou and A. Ortony (Eds.) *Similarity and analogical reasoning.* 498–531. Cambridge, MA: Cambridge University Press.

Strike, K. A., and Posner, G. J. (1985). A conceptual change view of learning and understanding. In L. H. T. West and A. L. Pines (Eds.), *Cognitive structure and conceptual change.* 211–311. Orlando, FL: Academic Press.

Thagard, P. (1992). Analogy, explanation, and education. *Journal of Research in Science Teaching.* 29, 537–544.

Thiele, R., and Treagust, D. (1991). Using analogies in secondary chemistry teaching. *Australian Science Teachers Journal.* 37, 10–14.

Thiele, R. B., and Treagust, D. F. (1994). An interpretive examination of high school chemistry teachers' analogical explanations. *Journal of Research in Science Teaching.* 31, 227–242.

Treagust, D. F. (1993). The evolution of an approach for using analogies in teaching and learning science. *Research in Science Education.* 23, 293–301.

Treagust, D. F., Harrison, A. G., and Venville, G. J. (1996). Using an analogical teaching approach to engender conceptual change. *International Journal of Science Education.* 18, 213–229.

Treagust, D. F., Harrison, A. G., and Venville, G. J. (1998). Teaching science effectively with analogies: An approach for preservice and inservice teacher education. *Journal of Science Teacher Education.* 9, 85–101.

Venville, G. J., and Treagust, D. F. (1996). The role of analogies in promoting conceptual change in biology. *Instructional Science.* 24, 295–320.

Venville, G. J., and Treagust, D. F. (1997). Analogies in biology education: A contentious issue. *The American Biology Teacher.* 59, 282–287.

Zeitoun, H. H. (1984). Teaching scientific analogies: A proposed model. *Research in Science and Technological Education.* 2, 107–125.

Zook, K. B. (1991). Effects of analogical processes on learning and misrepresentation. *Educational Psychology Review.* 3, 41–72.

Zook, K. B., and DiVesta, F. J. (1991). Instructional analogies and conceptual misrepresentations. *Journal of Educational Psychology.* 83, 246–252.

Zook, K. B., and Maier, J. M. (1994). Systematic analysis of variables that contribute to the formation of analogical misconceptions. *Journal of Educational Psychology.* 86, 589–699.

9

Solving Word Problems in Chemistry: Why Do Students Have Difficulties and What Can Be Done to Help?

Diane M. Bunce
Chemistry Department
The Catholic University of America

Abstract

Students often think that chemistry courses have "too much math". Most of this math is used to solve word problems that depict chemical situations. Teachers typically solve a sample problem or problems in class, assign additional problems for homework, and then put similar problems on tests. Many students have difficulty solving such problems on tests unless the problems are simple, one-concept problems that strongly resemble the sample problems from class or textbook. Teachers expect that if students have done their homework and thus practiced the simple problems, they should be able to apply this knowledge to more challenging problems in a test situation. Often the only real difference between the simple problems in class and the more challenging problems on the test are that the test problems have a different "story line" or combine two concepts into one problem. Such situations often baffle students. Many students solve problems "by analogy," which involves matching the new problem to those stored in long-term memory. All too often, the knowledge in students' long-term memory is not well organized; therefore, it is not easily accessible. Unlike experts, students tend to store solved problems in long-term memory according to surface features rather than according to the underlying chemical concepts. This type of long-term memory storage results in many students not being successful in solving problems and not knowing why.

There are ways to address this situation. The approach involves providing students with experience in problem analysis, not just problem solution. Students should experience each aspect of problem analysis with immediate feedback and support to help them overcome their natural inclination to immediately manipulate numbers to produce an answer. This chapter will discuss the causes of student failure with problem solving and some new approaches that teachers can use to help students become successful problem solvers.

Biography

I started my career in chemical education as a high school chemistry teacher. By the time I decided to go back to the University of Maryland to work on my Ph.D., I had taught high school chemistry in three states (New York, North Carolina, and Maryland). During my time as a high school teacher, I became aware of the fact that my students always seemed to have difficulty with the same topics in the course and most of these topics involved some sort of mathematically-based problem solving. I decided that I wanted to study student difficulty with problem solving in a more formal fashion, so I started a doctoral program with problem solving as the topic of my dissertation. Now, years later, I find that I am still interested in the mismatch between how we traditionally teach problem solving in chemistry and the way students learn. Writing this chapter has helped bring me back to where I started and the answers are still the same. To help students become successful problem solvers, we must help them understand and practice the different parts of problem analysis. It is not enough to show them that we, as experts, can solve these problems. We must construct an environment where students can analyze problems before computing the answers and thus build their own success.

Word Problems: What Are They?

Problems that are typically used for mathematically-based problem solving in chemistry are word problems that present a scenario and require a calculation to solve. The solving of word problems is recognized as being difficult for students. Students often read past the scenario, spend little time analyzing the problem, and immediately start entering numbers into an equation. An often-heard student comment is "I understand the problem; I just don't know whether to multiply or divide the numbers."

Word problems are found in close to 70% of the topics taught in a two-semester general chemistry course. The chemistry topics that typically include mathematically-based word problems are density, specific heat, stoichiometry, limiting reagent, equilibrium, empirical/molecular formulas, solution concentration (including molarity, normality, and formality), kinetics, acid-base calculations, solubility and dissociation constants, free energy, and electrochemistry including the Nernst equation. An example of a typical word problem is provided in Figure 1. Thus, it becomes obvious that problem-solving ability is something a student *must* master if he/she is to be successful in general chemistry.

Figure 1. Sample Word Problem (Chang, 1991, p. 57).

Phosphoric acid (H_3PO_4) is used in detergents, fertilizers, toothpastes, and carbonate beverages. Calculate the percent composition by mass of H, P, and O in this compound.

A more formal definition of word problems is that they define some given state and goal with no obvious or immediate method of getting from the former to the latter (Weaver and Kintsch, 1992). Another source (Kloosterman and Stage, 1992), defines a true problem as one that a person has no readily available procedure for solving.

Typical Student Approach to Solving Word Problems

Often, college students enrolled in chemistry courses believe that all mathematically-based problems can be completed in 5 minutes or less. They also conclude that if one cannot complete the problem within this time, he/she should give up on it (Kloosterman and Stage, 1992). This type of thinking runs counter to that of most professors, who expect that students will find some inherent satisfaction or even pleasure in being able to solve word problems and thus the time required to solve them is of no consequence. Some professors even go out of their way to include problems in the assigned list that will "challenge" their students.

Another belief about problem solving commonly held by students is that all word problems can be solved by applying the rules or equations presented in their textbooks. Little additional manipulation should be required. If students cannot readily find a rule to apply, many either give up or pick an inappropriate rule (Kloosterman and Stage, 1992). This blind application of rules without understanding how they work sets many students up for failure from the start. If students believe the main purpose of word problems is to find a rule, plug in the numbers, and then enter the appropriate numbers into their calculator, they are less likely to be able to solve the professor's "challenging" problems. Such students will also "cheat" themselves out of the opportunity to grow as successful problem solvers. Yet, by eliminating challenging problems from his/her teaching, a professor confirms students' misguided beliefs that all word problems can be solved by finding the appropriate equation and plugging in the numbers.

In research with deaf students (Mousley and Kelly, 1998), certain characteristics of unsuccessful problem solvers were identified including (1) impulsivity, (2) lack of transfer of knowledge from one situation to another, (3) breakdown in logical reasoning, (4) inability to organize and properly consider all relevant information in a problem solution, and (5) a misunderstanding of the problem goals. Although research with deaf students reports that students' impulsivity is much greater than that in hearing students, all five of these characteristics seem applicable to hearing students. Camacho and Good (1989) describe unsuccessful (hearing) problem solvers as those who have gaps in their knowledge base and numerous misconceptions. Their knowledge is fragmented and more difficult to access than that of experts. In addition, unsuccessful problem solvers often base their categorization of a problem on its surface features rather than the underlying concept. Combine this view of unsuccessful problem solvers with the idea that students think they must be able to solve a word problem within five minutes or less using one or two steps (Camacho and Good, 1989), and the differences between hearing and nonhearing students seem small.

Categorization of Problems by Surface Features. Students' general heuristic for solving word problems is to "solve by analogy." This means that they try to match the problem at hand to a solved problem either in their text or notes. Students persist in this behavior even when they have a complete set of step-by-step procedures available for solving specific types of problems (Reed, Willis, and Guarino, 1994). As a direct result of this problem-by-analogy solving technique, students will perform much better on test problems that have an "identical" solution rather than a "similar" solution to sample problems that they have seen in class or in their textbooks. Professors get a hint that this is how students are solving problems when students complain to the professor that they "studied all the problems in the text. They just couldn't solve the professor's problems." Students sometimes feel that the professor has made the test unfair by asking word problems that don't duplicate problems solved in class or in the text. This is a logical response on the part of the students *if* problem solving is limited to "solving by analogy."

There are other difficulties associated with students "solving by analogy". The greatest hindrance in using this approach to solve word problems is the effect of story context. When seeking analogies, students often choose sample problems that have similar story lines regardless of the fact that they may have different solution procedures. The use of these similar problems as analogies will result in incorrect answers. Sample problems that utilize the same solution procedures as the problem in question, but which have different story lines, are often overlooked by students when they search for analogies (Weaver and Kintsch, 1992). Figure 2 provides an example of two problems with different story lines, but similar solution procedures. This failure to use correct analogies based upon common concepts is a result of the students' behavior of selecting analogous sample problems based solely upon surface features.

Figure 2. Similar problems
(Different story lines, same solution procedures) (Weaver and Kintsch, 1992).

1. Train A leaves the station going east at 200 km/hour. Two hours later, Train B leaves the same station on a parallel track also going east, but going 250 km/hour. When will Train B overtake Train A?

2. A Girl Scout troop sells the same number of peanut butter cookies as chocolate chip cookies. Peanut butter cookies come 30 to a box, while chocolate chip cookies come 25 to a box. The troop sold 28 more boxes of chocolate chip than peanut butter. How many peanut butter cookies have been sold?

Even if students were to solve problem #1 correctly, they would not typically be able to apply the knowledge of how that problem was solved to problem #2. Students in this situation might think that "train" problems do not apply to "cookie" problems, even though the solution procedures of both problems are parallel. Experts recognize the two problems as being similar because they classify problems according to the underlying structure (Reed, Willis, and Guarino, 1994). Students classifying the two problems as train and cookie problems would not see the similarity between the two. This helps explain why the transfer of knowledge from one problem to a similar problem is so low for students. Verschaffel, De Corte, and Pauwels (1992) report successful transfer as occurring less than 20% of the time for novice problem solvers.

Translation of a Problem into an Internal Representation. In order to start solving a word problem, experts typically reinterpret the problem by constructing one or two qualitative representations before constructing a mathematical representation of the problem. These qualitative representations may include a pictorial depiction of the situation that defines the system at the start and finish of the process. The next representation is a physical depiction using diagrams and graphs that bridge the gap from the pictorial to the mathematical representation. The mathematical representation with equations or formulas is driven by the two qualitative representations of the problem. Students, on the other hand, often skip all qualitative representations and immediately search for a formula or equation to plug numbers into (VanHeuvelen, 1991). In a study comparing two introductory physics courses—non-calculus and calculus-based— (VanHeuvelen, 1991) only 10% of students in the non-calculus physics course used diagrams (physical representations) to solve problems, while 20% of the students in the calculus-based course used them. The reason why so few students draw diagrams to represent the problem before selecting an equation or formula is that most students do not understand the basic concepts represented in the problem. For instance, in physics problems, most students cannot successfully identify all the forces acting on an object even in simple situations. Up to 60% of those who "successfully" complete beginning physics courses are found to have basic misconceptions (VanHeuvelen, 1991).

The ability to represent problems qualitatively in a diagrammatic form may involve a student's visual-spatial aptitude or visual-spatial understanding (Trindade, Fiolhais, and Almeida, 2002). If students cannot visualize the problem in their minds, they will not be able to represent it in a qualitative format on paper. If they are not used to thinking visually, then problem solving may remain a search for the "correct" formula rather than an analysis of a given situation.

Many students have trouble just trying to understand what the problem is asking. The format of the word problem can add to this difficulty. Students expect problems to be written in a set form. For example, students expect the terms in a word problem to be consistent with the required arithmetic operation to be performed. Figure 3(a) presents an example of a word problem that is written in a manner consistent with the required arithmetic operation. If word problems are not written in such a manner, then the student must translate the problem before solving it so that it *is* consistent with the required arithmetic. This translation can result in errors before the problem's solution is even attempted. Figure 3(b) provides an example of a problem that is not consistent with the required arithmetic operation. The translation and subsequent solution necessary with an inconsistent word problem requires a good deal more memory space than the solution of a consistently written problem.

Figure 3. Examples of consistent and inconsistent word problems (Verschaffel, DeCorte, and Pauwels, 1992).

(a) Problem consistent with required arithmetic operation
Joe has 5 marbles. Tom has 3 more marbles than Joe. How many marbles does Tom have?
Analysis
The term "more than" in the second sentence is consistent with the required arithmetic operation of addition.

(b) Problem inconsistent with required arithmetic operation
Joe has 8 marbles. He has 5 marbles less than Tom. How many marbles does Tom have?

Analysis
The term "less than" in the second sentence implies the arithmetic operation of subtraction when addition is the required arithmetic operation.

Memory Demand of Problems
The actual memory demand of a problem can affect a student's ability to successfully solve it. Memory demand, or M-space, is similar to Piaget's belief that short-term memory is limited. According to Pascual-Leone's theory (in Roth 1992), short-term memory can handle about seven pieces of information. Only about 40% of college students are believed to have an M-space at the maximum of seven. Most are believed to have M-spaces of five or six. An average student then, can only successfully and reliably solve a problem that requires no more than five pieces of information. If the word problem demands more than the M-capacity of the student, the student will probably not be successful. The trickiest part of determining the M-space demand of a chemistry problem is that the number of pieces of information is dependent on the person's prior experience. If the problem solver has a good deal of experience with a particular concept, one or more pieces of information may be integrated and thus take up less M-space than they would individually. The professor, then, can use the concept of M-space to do an approximate check on word problems to make sure that, in their most elemental form, they do not exceed the memory of an average student. This is accomplished by doing a quick count of all the information provided in the problem that a student must deal with in the solution. For instance, a mass–mass stoichiometry problem requires conversion of mass of one substance to moles, then moles of this substance to moles of another substance, and finally conversion of moles of this second substance to mass. This type of problem has an M-demand of at least 3. If the concept of limiting reagent is combined with this type of problem, the M-demand will increase to at least 4. Limiting reagent or other concepts should not be combined

with mass–mass stoichiometry problems until the students have enough experience with mass–mass stoichiometry problems that the M-demand has decreased for the student. M-demand will decrease as the students become more familiar with the solution plan of a particular type of problem. Figure 4 includes an example of M- demands for typical problems.

Figure 4. Memory demands of typical problems (Chang, 2002, p. 89). All alkali metals react with water to produce hydrogen gas and the corresponding alkali metal hydroxide. A typical reaction is that between lithium and water:

$$2\,Li\,(s) + 2\,H_2O\,(l) \rightarrow 2\,LiOH\,(aq) + H_2\,(g)$$

How many grams of H_2 will be formed by the complete reaction of 80.57 g of Li with water?

Memory Demand
If the balanced equation is given, then the student must perform the following:
    1. **Convert grams of Li to moles of Li using the periodic table.**
    2. **Convert moles of Li to moles of H_2 using the balanced equation.**
    3. **Convert moles of H_2 to grams of H_2 using the periodic table.**

This results in a memory demand of at least 3. If the student is not familiar with the determination of the number of grams in a mole of an element or compound, the memory demand of the problem increases to 4 or 5.

Without a balanced equation, the memory demand increases substantially. In addition to completing the steps outlined above, the student would also be required to complete the following:
    1. **Recall what a metal hydroxide is.**
    2. **Balance the formula for LiOH.**
    3. **Recall that hydrogen is a diatomic gas.**
    4. **Balance the equation.**

These additional four steps increase the Memory demand to approximately 7, which is beyond the capacity of most students.

Student Use of Problem-solving Schema. Unsuccessful problem solvers usually make representational rather computational errors when solving problems (Verschaffel, DeCorte, and Pauwels, 1992). According to Riley (Zawaiza and Gerber, 1993), there are three critical types of knowledge that a problem solver must address when solving a problem. The first two, *Problem Schemata* and *Action Schemata,* are concerned with representation of the problem; the third, *Strategic Knowledge,* concerns computation. *Problem Schemata* involve understanding the relationship between the words of the problem and the underlying concepts. *Action Schemata* are primarily involved with planning the solution, while *Strategic Knowledge* includes the actual computation of the answer.

Problem schemata are the pictorial and physical representation of the problem referred to by Van Heuvelen (1991). Students tend to skip this qualitative analysis of the problem because they either don't fully understand the underlying concepts or they feel that this analysis is too time consuming. Part of this qualitative analysis involves identifying the data and variables relevant to the problem and relating them according to the conditions set in the problem (Adner, 1990).

Action Schemata are part of the problem analysis that divides the problem into sub-problems and coordinates the sub-problems into a unified plan of action. This plan of action moves the problem solver from the information given in the problem to the information requested. Selecting and manipulating the proper equation or formula is part of the *Action Schemata,* but this manipulation must be done in concert with the qualitative analysis of the problem from the *Problem Schemata*. It is only after all the analysis, planning, selecting, and manipulating are completed that the actual computation is attempted (*Strategic Knowledge*). Even if problem solvers analyze and plan the solution, computational errors are still possible. Computational errors are easier to correct than analysis or planning errors. Without proper Problem and Action Schemata, problem representation is often a trial-and-error approach, rather than an analysis, and the resulting problem understanding can be superficial (Zawaiza and Gerber, 1993). Figure 5 has an example of Problem Schemata, Action Schemata, and Strategic Knowledge demands for a typical problem.

Figure 5. Problem Schemata, Action Schemata, and Strategic Knowledge for a mass–mass stoichiometry problem (Chang, 2002, p. 89).

The following is a re-examination of the problem given in Figure 4 according to Problem Schemata, Action Schemata, and Strategic Knowledge.

All alkali metals react with water to produce hydrogen gas and the corresponding alkali metal hydroxide. A typical reaction is that between lithium and water:

$$2 \, Li \, (s) \; + \; 2 \, H_2O \, (l) \; \rightarrow \; 2 \, LiOH \, (aq) \; + H_2 \, (g)$$

How many grams of H_2 will be formed by the complete reaction of 80.57 g of Li with water?

Problem Schemata
Problem schemata present a pictorial or physical representation of the problem identifying data/variables and relating them to the conditions of the problem.

Given: 80.57 grams of Li
Asked for: grams of H_2
Recall: The number of grams of one mole of a substance can be determined from the periodic table
        1mole Li = 6.941 grams
        1 mole H_2 = 2.016 grams.

    The relationship between moles of two different substances can be determined from the balanced equation.
        2 moles of Li
        1 mole of H_2

Action Schemata
Action schemata divide the problem into sub-problems and coordinates sub-problems into a plan of action.

Plan:
        g of Li → moles of Li → moles of H_2 → grams of H_2

There are three steps outlined for the solution of this problem.

Strategic Knowledge
Strategic knowledge calls for the Computation of an answer:

$$80.57 \, g \, Li \; \times \; \frac{1 \, mole \, of \, Li}{6.941 grams \, Li} \; \times \; \frac{1 \, mole \, of \, H_2}{2 \, moles \, Li} \; \times \; \frac{2.016 \, grams \, of \, H_2}{1 \, mole \, H_2} = 11.70 \, grams \, of \; H_2$$

Problem solvers who do not understand the critical need for *Problem* and *Action Schemata* and who believe that computation or *Strategic Knowledge* is the only essential part of solving problems will have less motivation to strive to become good problem solvers (Kloosterman and Stage, 1992). These students believe that they are incapable of improving their math ability and thus will never be able to become better problem solvers.

Noncognitive Variables in Problem-solving. As vital as the cognitive variables are to successful problem solving, there are also some important noncognitive variables that can affect student success. These include a student's academic self-concept, achievement expectations, previous achievement, attitude towards problem solving, and ability to use logic (House, 1995). Students' verbalization of their potential success as problem solvers is the best method of gauging the impact of these noncognitive variables on a specific student's problem-solving success (Masui and DeCorte, 1999).

Unsuccessful problem solvers engage in a good deal of negative self-talk (Smith in Sugrue, 1993). This talk can include statements about the students being too slow or that their previous course work or experience in courses is inadequate to help them with the current course (Masui and DeCorte, 1999). In addition, unsuccessful problem solvers seem overly concerned with how many additional problems are left to be solved. They may also say that they feel uncomfortable trying to solve word problems and even that they dislike solving word problems or show a decidedly low interest in solving them (Camacho and Good, 1989).

Math and Test Anxiety. Students who have high math anxiety do not trust their own instincts and thus feel more comfortable solving word problems with someone more accomplished than themselves (Norwood, 1994). Such students are more interested in successfully determining the correct answer than understanding the *why* behind the solution. As a result, highly math-anxious students generally prefer to solve problems algorithmically rather than by analyzing them.

Students who experience test anxiety view themselves as inadequate or ineffective in problem solving. They expect to fail and worry about the loss of regard that the teacher or others will have for them when they do fail. It is estimated that 30% of all schoolchildren suffer from test anxiety and that 20% of this population will drop out of school because of it (Nicaise, 1995).

What Constitutes Successful Problem Solving?

After reviewing the components of problem solving, it may be helpful to compare the overall problem-solving procedures of novice versus expert problem solvers. Figure 6 describes the typical problem-solving approach of experts and novices.

Figure 6. Comparison of Expert and Novice's Approaches to Problem Solving (based on Camacho and Good, 1989).

Expert Problem Solver	Novice Problem Solver

1. Analysis
    a. Careful identification of variables and conditions in problem
	1. Little analysis performed
	    a. Poor translation of verbal problem into scientific language

2. Classification based upon underlying principles or concepts
    a. Knowledge of concepts organized and easily accessible
	2. Classification based on surface features (story line) of word problem.
	    a. Knowledge fragmented and difficult to access

3. High quality procedural and strategic knowledge a. Frequent checks of consistency of answers and logic	3. Little if any procedural or strategic knowledge

4. Solution plan based upon analysis of problem and conditions described in problem
    a. Flexibility in application of solution plan
	4. Solution plan based on use of pre-existing algorithm (formula, equation, prescribed set of procedures).
	    a. More guessing behaviors
	    b. Inflexible in application of solution plan.

5. Check of reasonableness of mathematical answer and procedure used to solve problem.	5. Neither checks of mathematical answer nor procedure normally done.

It should be noted that the distinctions cited here between expert and novice problem solvers represent opposite ends of the continuum. Students in most chemistry classes will be situated somewhere along this continuum.

What Can be Done to Help Students Become Successful Problem Solvers?

First, we must change students' perception of problem solving. Problem solving should be viewed as a dynamic interplay between the solver and the problem. This means that equations and formulas must be modified and/or combined to fit the conditions of the situation described in the word problem. Problem solving is not the static stuffing of numbers into a ready-made equation or formula taken off the shelf (Bernardo and Okagaki. 1994). The general strategy for successful problem solving includes (1) reading the problem, (2) reinterpreting the situation, (3) visualizing the interaction among variables, (3) hypothesizing on the underlying concepts involved, (4) computing the answer, and (5) checking to see that the answer is reasonable and that the procedures used to determine it are logical (Zawaiza and Gerber, 1993).

What we do know from research on problem solving is that internalization and application of the skills demonstrated by experts can be learned by novices through (1) repeated practice, (2) active participation, (3) interactive discussion, and (4) evaluative feedback (Mousley and Kelly, 1998). Telling students the needed concepts and techniques to successfully solve a problem has proven to be ineffective. Even though this transmission of the information is efficient, the understanding gained by students is almost negligible (VanHeuvelen, 1991).

The traditional approaches to teaching problem solving have included teacher lectures, followed by practice solving problems for homework, and possibly reviewing the correct solution in class by having students write solutions on the board. The results of this teaching approach are that unsuccessful problem solvers are still unaware of what to do to solve the problem. They can't think of options or overcome the fixation on one possible solution, even if it is not working. They often misread the problem and either miss or misinterpret the important variables and conditions given in the problem (Woods, 1993). What's missing from this approach to problem solving is the explicit, prompt, and personal feedback on each of the sub-skills necessary to solving problems. These subskills include (1) visualizing and constructing a diagram of the problem situation, (2) identifying the underlying concepts/principles of the problem, (3) developing a solution plan, (4) computing the answer, and 5) checking the answer and solution plan for accuracy and logic. There are alternative teaching techniques that will help students acquire these subskills.

1. Visualizing and constructing a diagram of the problem situation. Zawaiza and Gerber (1993) showed that training students to develop diagrams of the problem situation as a way to identify the components of the problem and to physically represent the relations between the components caused a significant increase in the students' problem-solving success and had a lasting effect on students even after the formal training was finished. Such training showed students how the components of the problem are related to each other and was significantly more effective in increasing students' problem-solving success than either instruction in translating the problem into a format that was consistent with the math operation to be performed or providing opportunities for students to discuss their individual solution strategies. Such diagrams do not need to be elaborate, but they must identify the pertinent components of the problem situation and show how they are related to each other. Working within mixed ability groups in class can help unsuccessful problem solvers develop a more systematic approach to developing these diagrams. It can also help unsuccessful problem solvers analyze both their own approach and those of more successful problem solvers (Mousley and Kelly, 1998). Emphasis on diagramming a problem before starting the actual solution will help inhibit students' natural tendency to jump into problem solving without thinking first. In chemistry, an effective diagramming strategy often involves the construction of a particulate diagram that depicts the interaction between atoms, ions, or molecules. Such diagrams can be constructed for problems within mixed-ability small groups in class, and the groups' results can be discussed with the whole class. This would help nonsuccessful problem solvers see different approaches used by their peers to diagramming problems. Figure 7 provides a sample particulate diagram for a typical stoichiometry problem.

Figure 7. Visualizing and constructing a diagram of the problem situation.

Using the same mass–mass stoichiometry problem from Figures 4 and 5 (Chang, 2002, p. 89), the visualization of the problem would involve the interaction between Li atoms and H_2O molecules to form the metal hydroxide LiOH and hydrogen gas H_2.

All alkali metals react with water to produce hydrogen gas and the corresponding alkali metal hydroxide. A typical reaction is that between lithium and water.

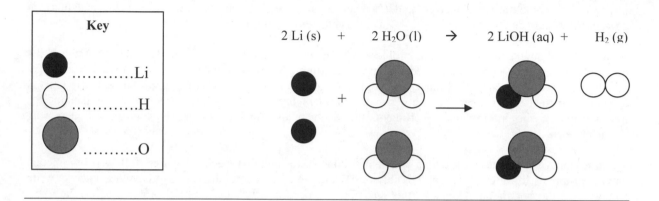

2. Identifying the underlying concepts/principles of the problem. It should not be assumed that students recognize or understand the underlying concepts of a problem. Opportunities should be provided either through group work or through one-on-one interaction with a teacher/mentor for the student to identify and explain the concepts involved in a problem's solution. The most important part of this process is that the student is doing the analysis, yet has access to immediate feedback from either members of the group or the teacher/mentor. This step in the problem-solving procedure should be explicitly practiced by the student before attempts are made to plug numbers into the calculator. One benefit of taking the time to practice this aspect of problem solving is that students should be able to develop a more integrated knowledge structure which will make access of their knowledge easier and more efficient.

3. Developing a solution plan. Developing a solution plan is the third step in the analysis process that does not come naturally to unsuccessful problem solvers. Rarely do they take the time to think about how they will get from what is given in the problem to what they are asked for. Such a procedure does not naturally fit in with their five-minute rule of problem solving. Even though expert problem solvers may not be fully aware of the fact that they have planned a solution, if you stop them during their solution process, they will most likely be able to tell you what they intend to do next in their solution. Unsuccessful problem solvers will not. Practicing the development of a solution plan either individually or in a group helps the novice problem solver test the logic of his/her approach from the beginning to the end of the problem solution before the situation is clouded with numbers and mathematical manipulations. Such a solution plan does not need to be elaborate. A simple flowchart-type diagram that proceeds from the "given" to the "asked for" would suffice. The solution plan is also a way to utilize the concepts that have been previously identified by the student as being important in the solution.

4. Computing the answer. If the analysis and solution plan have been developed, then the computation part of the solution is trivial. With a solution plan that is written out, any errors that appear in the computation can be easily identified as either trivial or central to an understanding of the problem. Grading in such a manner helps the student understand that multiplying two numbers together incorrectly is not as important as not knowing why they should be multiplied in the first place. Grading emphasis is shifted to reflect the relative importance of problem analysis over problem computation.

5. Checking the answer and solution plan for accuracy and logic. Students can easily be convinced that they should spend time checking the accuracy of their math calculations, but few would naturally think to re-examine their solution plan for accuracy and logic. However, by not taking the time to review their solution plan, many students will not be able to access the success they had analyzing and solving a particular problem when they solve a similar problem (i.e., which has the same underlying concepts and a different story line). By taking a few minutes to reread the original problem and solution plan, students will be better able to see the underlying concepts and how they were used to solve the situation depicted in a particular story line. As a result, the particular problem-solving experience is more likely to be stored in long-term memory according to concept rather than story line.

The role of the teacher in problem solving, then, is not merely to demonstrate how to solve problems, but rather to provide students with a guided opportunity to solve the problems themselves. The teacher can manipulate the environment by providing students with the support they need by having students work on problems in groups. As important as group problem solving is, it is not sufficient. The teacher must also break the problem-solving process down into its component parts as discussed here, identify objectives for each part and provide feedback for student effort on each part. This action is necessary to break students of their inclination to breeze through word problems with their hands on the keys of the calculator to produce a single numerical answer. Such calculator action fits the five-minute approach to problem solving and results in many students being unsuccessful without understanding why they are unsuccesful. Teaching the component skills of the problem-solving process with increased emphasis on problem analysis will help students become successful problem solvers and appreciate the mental exercise such problems afford the learner.

Suggested Reading

Woods, D. R. (September/October 1993). Problem solving-What doesn't seem to work. *Journal of College Science Teaching.* 57-58.

Provides a simple overview of problem solving and the difficulties exhibited by students.

References

Adner, H. (1990). Algebraic versus programming approach to solving mathematical modeling problems. *School Science and Mathematics.* 90(4): 302–316.

Bernardo, A. B. I., and Okagaki, L. (1994). Roles of symbolic knowledge and problem-information context in solving word problems. *Journal of Educational Psychology.* 86(2): 212–220.

Camacho, M., and Good, R. (1989). Problem solving and chemical equilibrium: Successful versus unsuccessful performance. *Journal of Research in Science Teaching.* 26(3): 251–272.

Chang, R. (1991). *Chemistry.* New York: McGraw-Hill, Inc.

House, J. D. (1995). Noncognitive predictors of achievement in introductory college chemistry. *Research in Higher Education.* 36(4): 473–490.

Kloosterman, P., and Stage, F. K. (1992). Measuring beliefs about mathematical problem solving. *School Science and Mathematics.* 92(3): 109–115.

Masui, C., and DeCorte, E. (1999). Enhancing learning and problem-solving skills: Orienting and self judging, two powerful and trainable learning tools. *Learning and Instruction.* 9: 517–542.

Mousley, K., and Kelly, R. R. (1998). Problem-solving strategies for teaching mathematics to deaf students. *American Annals of the Deaf.* 143(4): 325–336.

Nicaise, M. (1995). Treating test anxiety: A review of three approaches. *Teacher Education and Practice.* 11(1): 65–81.

Norwood, K. S. (1994). The effect of instructional approach on mathematics anxiety and achievement.*School Science and Mathematics*. 94(5): 248–254.

Reed, S. K., Willis, D., and Guarino, J. (1994). Selecting examples for solving word problems. *Journal of Educational Psychology*. 86(3): 380–388.

Roth, W.–M. (February 1992). How to help students overcome memory limitations. *Journal of College Science Teaching*. 210–213.

Sugrue, B. (1993). Project 2.1 Designs for assessing individual and group problem-solving: Specifications for the design of problem-solving assessments in science. Los Angeles, National Center for Research on Evaluation, Standards, and Student Testing: 1–58.

Trindade, J., Fiolhais, C., and Almeida, L.. (2002). Science learning in virtual environments: A descriptive study. *British Journal of Educational Technology*. 33(4): 471–488.

VanHeuvelen, A. (1991). Learning to think like a physicist: A review of research-based instructional strategies. *American Journal of Physics*. 59(10): 891–897.

Verschaffel, L., DeCorte, E., and Pauwels, A.. (1992). Solving compare problems: An eye movement test of Lewis and Mayer's consistency hypothesis. *Journal of Educational Psychology*. 84(1): 85–94.

Weaver, C. A. and Kintsch, W.. (1992). Enhancing students' comprehension of the conceptual structure of algebra word problems. *Journal of Educational Psychology*. 84(4): 419–428.

Woods, D. R. (September/October 1993). Problem solving—What doesn't seem to work. *Journal of College Science Teaching*. 57–58.

Zawaiza, T. R. W., and Gerber M. M. (1993). Effects of explicit instruction on math word-problem solving by community college students with learning disabilities. *Learning Disability Quarterly*. 16: 64–79.

10

An Introduction to Small-Group Learning

Melanie M. Cooper
Department of Chemistry
Clemson University

Abstract

There is increasing evidence that the traditional 50-minute science lecture (even with the use of lecture demonstrations and PowerPoint presentations) and accompanying cookbook laboratories are not effective for a majority of our students. Much of this book is devoted to why this is so, and to alternative ways to enhance the opportunities for students to learn in a meaningful way. One such method, group learning, has become increasingly popular in college-level science courses over the past few years. Although there is a vast amount of research into the use and effectiveness of small groups, most of it was based on the K–12 classroom. There is now an increasing amount of evidence that the use of small groups in college-level classrooms can lead to a more effective learning environment and increased student satisfaction. There are a number of ways that small-group learning can be implemented all through the chemistry curriculum from introductory general chemistry laboratories and lectures to upper-level problem-solving courses. It is important, however, to be aware of the potential pitfalls of small-group learning, and the most effective uses of the various ways that groups can be incorporated into your classroom.

Biography

Melanie Cooper is currently the Alumni Distinguished Professor of Chemistry at Clemson University. She has taught organic and general chemistry for the last seventeen years, including large enrollment lecture classes, laboratory sections, and small discussion classes. She is committed to the use of active learning techniques including small-group learning in her classrooms and labs, and has authored a laboratory manual that makes extensive use of cooperative learning groups in the general chemistry laboratory.

She has served as the principal investigator on a number of educational research and curriculum development projects funded by the National Science Foundation and the U.S. Department of Education on such topics as project-based laboratories, problem solving, and web-based resources for chemistry education. She is also a member of the editorial/writing team for the ACS "Chemistry" project, a new general chemistry curriculum that is based upon inquiry and active learning techniques.

Introduction

There is plenty of evidence (see Chapters 1 and 2) that the traditional methods used for teaching chemistry at the college level, including the instructor-delivered lecture, the verification laboratory, and the teaching assistant–led recitation session (where students are shown how to solve problems) are not effective. Most educators now believe that "telling is not teaching" and that having students verify concepts and facts about which they are well aware is at best wasteful of time and resources and at worst counterproductive. How many of us have despaired, after giving a wonderful lucid lecture, only to be faced with questions in class and answers on examinations that indicate a lack of understanding on the part of the recipients of this lecture? Despite clear evidence from the researchers and personal evidence from our students that traditional methods of instruction are not working for many students, we often continue to teach this way and yet each year expect a different result. It might be fair to say that most of us do an excellent (!) job of teaching chemistry, but the trouble lies in the way we educate chemists.

Over the past few years, as we have come to a better understanding of how students learn (NRC 2003), there have been increasing calls for science educators to incorporate researched-based teaching methods into their classrooms and laboratories (NSF 1996, NRC 1997). One of these methods (or several, depending upon which definitions you use) calls for the use of student groups or teams to solve problems, conduct laboratory experiments, or learn specific material. Obviously, one of the outcomes of using student teams should be that students learn the value of teamwork and how to work together. This skill has been recognized as important at least since 1945 when the ACS Committee on Professional Training called for the incorporation of teamwork into chemistry curricula. As we shall see, allowing students to work together results in a wide range of benefits, both in student learning and other less tangible results.

This introduction to small-group learning will range from brief activities during lectures to the use of long-term structured cooperative learning work. Since another chapter in this volume will focus on the out-of-class learning method known as Peer-Led Team Learning (PLTL), this method and other out-of-class group learning activities, such as Supplementary Instruction (SI), will not be discussed here.

Terminology

The terms *cooperative learning*, *collaborative learning*, *small-group learning*, and *team-based learning* are often used interchangeably by practitioners, although many educational researchers do draw distinctions among these various terms. Cooperative learning most commonly denotes the use of well-structured long-term groups, where each member has a specific role to contribute to a well-defined learning task; collaborative learning and team learning can encompass a much wider range of activities from structured to informal. A taxonomy proposed by Cueso (1992) places cooperative learning as a sub-type of collaborative learning. However, a more practical approach may be to think of collaborative and cooperative learning as the ends of a continuum, with collaborative learning at the unstructured end and cooperative learning at the structured end. As we will see, many practitioners move freely along this continuum, employing the most appropriate technique for the task at hand. In practice, I will try to make clear the extent and type of the collaborative groups as I discuss them.

NAME	CHARACTERISTICS OF GROUP	TYPICAL USE
Small-Group Learning	All encompassing term, can be used to describe any kind of group activity. Group can be from two to five members, short term or long term.	All types of group work, laboratory, lecture discussion, problem solving
Collaborative Learning	All encompassing term, can be used to describe any kind of group activity. Group can be from two to five members, short term or long term.	All types of group work, laboratory, lecture discussion, problem solving

Cooperative Learning	May either be considered as a subset of collaborative learning, or as one end of the collaborative learning spectrum. Groups are typically long term with three to four members, and assigned roles.	Cooperative learning is often appropriate for laboratories where each member may have a specific role, and for long-term projects where group members are responsible for teaching and learning from each other
Peer-Led Team Learning (PLTL)	Long-term group with student leader/coach. Typically up to 10 members.	Out-of-class problem-solving sessions.

History

The history of group learning is long, is well researched, and can be traced back to the archetypal one-room schools where, in vertically structured learning groups, older students taught the younger ones. This history continues through the spontaneous formation of informal study groups in many courses to the present day, where students may participate in a wide variety of group activities in a broad spectrum of courses.

In the past 100 years, there have been numerous research studies on the effects of using small student groups, although most of the research has been focused on pre-college students (Slavin, 1996). This is because the formal use of cooperative and collaborative learning has its beginnings in elementary education settings and has moved up through the curriculum as the techniques have been shown to be effective. In the past 15 years, there have been an increasingly large number of studies and reports on the effect of collaborative groups on higher education student learning in chemistry. Probably the most influential work on cooperative leaning in higher education resulted from the research of Johnson, Johnson, and Smith (1998), and their book *Active Learning: Cooperation in the College Classroom* is a valuable resource for practitioners of any type of group work. Another resource also targeted specifically to the needs of higher education is also available (Millis and Cottell, 1998). Both of these works, however, are aimed at a fairly general audience (rather than science instructors), although both have some science specific resource material.

Does Small-Group Learning Work?

Clearly, before adopting any unfamiliar educational technique (or a familiar one, for that matter), we should begin by asking if there is any evidence that the technique is effective and that it will have a positive impact on student learning and the classroom environment. Fortunately, there is a huge database of research on small-group learning that is *"longstanding and solid"* (Mills, 1998). As Johnson and Johnson (1991) reported,

> *"we know more about the efficacy of cooperative learning than we do about lecturing ...... or almost any other facet of education."*

It is now well established that group work can have significant positive effects on many aspects of student learning, including achievement, attitudes, and persistence (Slavin, 1996). The research-based outcomes according to Johnson and Johnson are

- higher achievement.
- increased retention.
- more frequent higher level reasoning, deeper level understanding, and critical thinking.
- more on-task and less disruptive behavior.
- greater intrinsic motivation to learn.
- greater ability to view situations from others' perspectives.
- more acceptance of diversity.

- greater social support.

- more positive attitudes towards instructors.

- more positive attitudes towards learning.

- greater social competencies.

As chemistry instructors, we tend to focus on the areas of achievement as measured on tests, but we should be aware that the other positive outcomes listed here are vital to the overall experience of the student. In fact, Astin's landmark study (1993) of student achievement over 20 years indicated that there were only two factors that were always positively correlated with student achievement: student–student interactions and student–instructor interactions. In addition, numerous studies for potential employees typically rank the "soft" skills such as communication (both oral and written), teamwork, and leadership as more important than the technical skills and content that we typically concentrate on in chemistry courses. For example, the National Association of Colleges and Employers (NACE) reported in their Job Outlook 2003 Survey that all these social skills were highly valued by employers.

Until relatively recently, there was not a large body of research into small-group learning at the college level in the science and technology disciplines. In 1998, a meta analysis conducted by Springer et al. analyzed the results of 39 reports on small-group learning dating from 1980 which presented data on achievement, retention, or attitudes. Their analyses let to the conclusion that

> *"various forms of small group learning are effective in promoting greater academic achievement, more favorable attitudes towards learning, and increased persistence through SMET (science, math, engineering and technology) courses and programs. The magnitude of the effects reported in this study exceeds most findings in comparable reviews of research on educational innovations and supports a more widespread implementation of small-group learning in undergraduate SMET."*

The authors also reported that out-of-class group work, such as supplementary instruction, was more effective at increasing achievement than in-class group work, but in-class work was more effective at increasing student satisfaction with the course. In this analysis, however, only three of the reports were based on chemistry education.

A similar meta analysis by Bowen (2000) looked at 15 studies of cooperative learning outcomes in chemistry alone. Again, the majority of the reports indicated a positive effect on student learning and the author recommended that cooperative learning practices be incorporated into chemistry classrooms. It is important to note that both of these reports (Springer, 1997; Bowen 2000) included all types of small-group work in their analysis, not just formal structured cooperative learning. Neither made any real distinction among cooperative, collaborative, or small-group activities and programs.

Examples of Small-Group Work in Chemistry Courses

Collaborative groups have been used in a wide variety of situations ranging from large enrollment lecture classes to upper-level laboratory courses. There are a number of annotated bibliographies with examples at all levels; see Cooper and Robinson (1998) for all of the STEM disciplines, and see Robinson and Nurrenbern (1997) for chemistry.

A.   In the Laboratory

Probably the place where small-group activities have become most accepted is in the laboratory. Analytical chemistry has been in the forefront of the use of groups in the laboratory, probably because of its relatively close ties with industry, where the value of teamwork has long been known. One of the earliest and most influential reports of the use of groups in a lab is that of Walters (1991), who pioneered the use of structured groups in the analytical laboratory. Each student assumed a different role (manager, hardware, software, or chemist), and the group approached an analytical problem as if they were an industrial team. According to Walters,

> *"One of the advantages of role playing in the Chemistry laboratory is to allow the development of technical expertise at an individual level while at the same time stressing and developing communication and collaborative skills. It is particularly effective for teaching undergraduate Analytical Chemistry because much Analytical work naturally thrives in an interdependent, diverse small group."*

Small-group laboratory programs are now relatively commonplace in all areas of chemistry, from introductory general chemistry lab programs with thousands of students to upper-level physical chemistry laboratories, where the groups may not even be on the same campus (Towns, 2001). There are now a large number of laboratory exercises reported in the *Journal of Chemical Education* in which students may work in teams to solve the problem at hand, and a number of laboratory manuals where cooperative learning is explicitly incorporated into the nature of the laboratory (Birk and Bauer, 2001; Cooper, 2003). A resource from the Institute for Chemical Education (Nurrenbern, 1995) contains collections of appropriate laboratories for small-group activities. In addition, guided inquiry laboratory programs (Abraham and Pavelich, 2004) and activities, while not explicitly requiring small groups to be effective, are often done this way, for reasons we will discuss later in the chapter (see *Why does small group learning work?*).

One (of the many) advantages of using structured cooperative learning groups in the laboratory is that students can be given a more complex problem to solve than they might otherwise receive. It is almost always true that the discussion and reflection that students must engage in to solve a complex problem will produce a more thoughtful and successful solution than most students could come up with alone. Therefore, many laboratories are now beginning to incorporate real-world problems and indeed may even be listed as offering problem-solving courses (Wilson, 2003; Wink, 2000; see also Chapter 5 for a discussion of relevance in the curriculum).

B.   In the Lecture

It has been established by a number of researchers using a number of methods (observations, heart rate monitoring, quantity and quality of note taking, and recall of material) (Bligh, 2000; MacManaway, 1970; Scerbo, 1992) that, by 25 minutes into a lecture, most students have severe lapses in attention, and material presented after this time is poorly understood by students. One way to counter this problem is to incorporate small-group activities into the lecture at these vulnerable times. The change in pace brings students back to the subject of the lecture and increases their attention.

The use of small-group learning in lecture or discussion classes has again been well documented. Once more, there is a continuum of experience with small groups, from informal pairing during lecture to whole courses taught in a structured cooperative learning mode.

At the Collaborative Learning end of the spectrum, the use of "think pair share" was first proposed by Lyman (1981). In this technique, students are posed a question, asked to think about it, then asked to discuss it with a neighbor. Depending on the size of the classroom, the instructor can either then solicit answers from the students or take a vote. The technique of voting answers is very useful for large classes where even the most outgoing students are sometimes intimidated and reluctant to speak up. In addition, the "pair" discussion ensures that every student in the class has the opportunity to think over the question and engage actively in the class. Another advantage of this kind of activity is that it is low risk—not much time is taken out from the lecture, and, as such, it is an ideal introduction for instructors who want to begin introducing active learning techniques into their repertoire.

The use of what has come to be known as ConcepTests (Mazur, 1997) is an outgrowth of this kind of collaborative technique, particularly in large enrollment lecture courses. Students are challenged with a question that tests their understanding of a particular concept, asked to think about it, and asked to answer the question individually. Students then discuss the question with a partner or group and answer the question again. Mazur's studies of students in an introductory physics class showed that a statistically significant improvement in student performance was obtained when these ConcepTests were included as part of a traditional lecture format course.

These kinds of activities are an ideal introduction to using small-group learning precisely because they do not require much investment of time or effort on the part of the instructor, yet they have been shown to increase student performance in some situations.

Other variations on this theme of short student-centered activities in lecture include giving group quizzes (Cooper 1995) and short inquiry-based learning activities (Wright, 1996). A short group activity is an ideal break in the lecture and will allow students to refocus on the ideas being discussed. There is ample evidence that, despite what may seem to be a riveting lecture delivery on the part of the instructor, students' attention seems to decrease as the lecture proceeds. (See Bligh, 2000, for an excellent discussion of factors affecting student attention.)

There are now a number of reports of courses that have been almost totally converted to group activities and inquiries. A complete general chemistry course has been designed (Farrell, Moog, and Spencer, 1991) where students in groups of four work through guided inquiry worksheets to develop and understand concepts. This process-oriented guided inquiry (POGIL) approach has been extended to a technological approach LUCID (Hanson and Wolfskill, 2001), where groups of students use computer activities as inquiry-based learning tools. For introductory physics classes, Meltzer (2002) has developed a set of criteria for the format, procedures, and curriculum materials for a totally interactive classroom where small groups are the norm.

In other chemistry courses, small groups are also becoming more common; for example, Maitland Jones reported on his experiences in *Teaching the Sophomore Organic Course without a Lecture. Are You Crazy?* (Bradley, 2002). An assessment study of student achievement in the course found no difference between the group problem-solving and lecture-based courses, but student satisfaction and retention are much higher for the non-lecture courses. This theme is also reported by Williamson (2002) in a group-based problem-solving analytical course. Again, no gains in achievement were made, but significant increases in satisfaction and higher retention rates were observed.

C.   In the Recitation or Discussion Session

Many introductory courses are accompanied by a recitation or discussion session, where a variety of activities take place. Traditionally, these sessions have focused on problem solving, and there is often a short quiz during the session. The sessions are an ideal place to incorporate small-group activities, and there are a number of notable successes in this area. Many of the guided inquiry materials developed by the POGIL (http://www.pogil.org/) organization are appropriate for recitation sessions for those institutions where it is neither possible nor desirable to convert the whole course to guided inquiry. McDermott (2001) has used her research into student misconceptions of scientific concepts (Vokos et al., 2000) to develop a set of materials for physics—*Tutorials for Introductory Physics*—that should serve as a model for materials of this kind. These tutorials, guided by the insights that the author has gained from investigation of how students learn physics, allow students to work in groups through materials designed to elicit understanding and confront misconceptions.

There is, however, a strong caveat that must be heeded when using recitation sessions; many institutions have teaching assistants who lead these sessions. If the teaching assistants, or whoever leads the recitations, do not "buy in" to the ideas of cooperative learning and active involvement, then these methods will be no more successful than traditional recitations where the leader solves problems for students. This, in turn, will lead to dissatisfaction on the part of both students and session leaders. This situation is often observed when new teaching materials are introduced at large universities. As we know, chemistry educators are a conservative lot, and their graduate students are no different. A dedicated program of TA training can overcome the innate resistance to change that can often be found in large institutions. For example, the ChemTA program of Chemistry Teaching Assistant Professional Development at Berkeley (http://www.ocf.berkeley.edu/~chemta/) provides materials to support teaching assistant development and active learning materials for discussion sessions.

Why Does Small-Group Learning Work?

There can be little doubt that small-group learning results in increased student achievement on tests, increased retention in courses, and increased student satisfaction with their educational experiences. However, there is much less research about *why* small group learning is so effective. One obvious possibility is that students in groups simply spend more "time on task." Springer's meta analysis indicates that out-of-class supplementary instruction groups are more effective than in-class groups in improving student performance on tests. Simply spending more time on the subject, however, cannot be the only reason why groups are so effective in so many different arenas. For possible explanations, we need to turn to modern learning theories (see Chapter 1 for more details):

- **Constructivism:** We now know that knowledge and skills, such as problem solving, are not transferred intact from the instructor to the student—even though we still spend a great deal of time teaching as if this were true. In fact, for material such as chemistry to be retained and understood at a deep conceptual level, the student must internalize the material and make it their own. This concept is commonly known as constructivism (Wheatley, 1991; Bodner, 2001). Students working in groups have the opportunity to grapple with difficult concepts and face the fact that they do not understand something. This leads us to metacognition.

- **Metacognition:** This term literally means "thinking about thinking." Studies have shown that asking students to think about how they will solve a problem or proceed with a task and then asking them to review their thought processes after the task (metacognition) improves student performance on that task (Rickey and Stacey, 2000). In a cooperative group, students must not only think about what they are doing and why, but they must also verbalize the same information and defend their strategy to their teammates. This may lead to a deeper understanding and also allow the student to grapple with concepts and problems that they might otherwise give up on. The very fact that students must communicate their thoughts to someone else may lead to cognitive dissonance.

- **Cognitive dissonance**: Many students are hampered by misconceptions about science, which can be very difficult to dislodge. Misconceptions are often resistant to instruction, as evidenced by the entering graduate students at a major university who held a wide range of alternate conceptions of common chemical concepts and principles (Bodner, 1991; Mulford and Robinson, 2002). It is often necessary to provide an experience that will require the student to directly confront their misconception—a so-called discrepant event—which cannot be explained on the basis of the student's current understanding. This induces a disequilibrium or cognitive dissonance (Kubli, 1983; Kuhn, 1981; Perret–Clermoat, 1980) in which the student must confront the misconception and construct a new understanding. Working in a group situation may result in a more frequent identification of these misconceptions and may also give students the structure and support that will allow them to delve into misconceptions; this is intrinsically uncomfortable.

The investigation of just how and when these principles come into play in collaborative groups is one that remains to be explored in more detail.

How To Get Started With Small-Group Learning

Preparing students to work in teams can be as simple as pairing them up randomly to discuss a question or compare answers. However, for more complex assignments, or for team projects of longer duration, the structure provided by a formal cooperative learning group will certainly prove more effective than simply sending a group of students off to tackle a complex problem or project. In fact, most of the perceived problems associated with having students work in teams can be traced to insufficient preparation of the group to work together (Fiechter and Davis, 1985). Most experts agree that

> *"putting students into groups to learn is not the same thing as structuring cooperation between students"*

(Johnson, Johnson, and Smith, 1991). Therefore, if you decide to use small-group learning for anything more than short lecture breaks, it is important to structure the groups properly.

The most important elements that contribute to the success of cooperative learning are *positive interdependence* and *individual accountability*. That is, for any group activity to succeed, the group must see that each person in the group benefits from the others' efforts—that they must cooperate rather than compete—or sink or swim together. Each member must also realize that ultimately the individual members of the group are accountable and it is the effort of the individual that will be rewarded in the long term.

To prepare a formal cooperative learning group to work effectively, several factors should be considered:

Team make-up: Most college level practitioners prefer heterogeneous groups of four students for structured groups for several reasons, chiefly because the groups are small enough so that no one can hide, but large enough to function should a member be absent. Four students also can work in pairs if necessary—another effective technique for less structured assignments.

Group roles: In a cooperative learning group, students are often assigned roles within the group to ensure that each member knows what their responsibilities are to the group and within the project as a whole. Typical roles might be leader, reporter, facilitator, and record keeper. Each role has responsibilities that are agreed upon before the project, and these roles are often rotated in long-term groups to promote interdependence. In laboratories, roles can be chosen that fit in with the nature of the assigned task; for example, in an analytical chemistry lab, students might take on the role of group leader, technician, computer analyst, wet chemist, etc.

Team building: For long term cooperative groups, team building is an important—but often omitted — contributor to the success of the team. Team building exercises structure group cohesion, and allow group members to develop trust and begin to know each other. For collaborative groups, however, this step is not as important since typically the stakes are not as high in a short-term group. Team building exercises are often divided into

- ice breakers—used to help teams get acquainted
- role assignments—used to reinforce the responsibilities of each student
- open-ended problems—used to help students learn to work together on a low-risk assignment.

There are a fairly large number of resources available to help instructors build effective teams (Kagan, 1992; Towns, 1998; Felder, 2001) and deal with any assorted problems that may arise.

Potential Problems

No discussion of small group learning would be complete without at least acknowledging that there are a number of potential (real or perceived) problems that can arise when students work in groups.

A.   The Hitchhiker Problem

One fear of many who are new to Cooperative Learning is that the "good" students will resent working with others who may not have their own aptitude or work habits, and who might "hitchhike" onto the good students' work. There are a number of ways to combat this perception:

- Don't allocate a large portion of the class grade to any group activity; any grade the student obtains should come mainly from individual effort, not from the group. "Good" students can see the hitchhiker will not be rewarded just for "being there." For example, some instructors do not employ group quizzes in class because they think the "poor" students will copy from the "good" students. While this may be true in a few cases, most students work to solve the problem in good faith. Student comments such as

> *"group quizzes made me prepare before coming to class because I didn't want to look bad in front of my group mates"*

   are more indicative of student attitudes to these quizzes (Cooper, 1995) Another approach to encourage group collaboration can be used after a traditional exam. When students have submitted their answers, they are given another shorter period to re-do the exam as a group. If the group makes a grade higher than that of the highest individual, then everyone in the group receives a bonus (usually five points). This type of group activity is an excellent illustration of the benefits of working together.

- Don't grade on a curve; this will encourage competition, not collaboration. If students know that only a few of them will succeed (make an A), they are less likely to help each other. (For more information on effective grading, see Greenbowe and Burke, 2003)

- Make sure that students understand the rules regarding when they can collaborate and what the benefits are likely to be. For example, if students understand that the purpose of group quizzes is to help them learn the material, rather than to assess them, they will be much more comfortable helping each other. In laboratories, where the problem to be solved is more involved than a five- minute calculation, it is more effective to use structured cooperative groups. Each student can be given a specific task and must understand his or her role and contribution to completing the task. Simply giving the whole group a problem to solve will often result in three students watching the fourth conduct an experiment.

B.   The Coverage Problem

Many instructors see the need to "cover the material" as the major drawback to using class time for small-group activities. Since incorporating group activities into a class will take away from lecture time, it would seem that less material can be "covered". However, there is no evidence that the use of small groups has a detrimental effect on student achievement, rather the effects are almost always reported to be positive. As we learn more about how students learn, it should become clear that the best use of our time is probably not by clearly explicating the contents of the book and demonstrating how to do problems, but rather by providing an environment where students can learn in a variety of ways including working with others to learn and solve problems.

To solve the problem of the effective use of class time, students can use a number of methods. Some instructors use a mixture of pre-class web-based quizzes and in-class collaborative learning activities—an approach known as "just-in-time" learning, first introduced into physics classes (Novak, et al, 1999). Others use assigned readings before class, and then the class time can be used to clarify, discuss, and work problems. Both the instructor and the students will be engaged and working hard (rather than just the instructor). It is probably a mistake to think that even though we are lecturing with great enthusiasm and clarity, we have control over what students are thinking and learning.

C.   The Control Problem

This may not seem like a problem until you have 200 students all vigorously discussing chemistry. Then it can be quite daunting to try to regain order. It pays to make it clear to the students what the signal will be for the end of the discussion phase. Students learn quickly and soon pick up on your signals to close discussions. It is possible to do group work even in a large class and regain control of the class, although you should be aware that the class atmosphere will probably be significantly different from a formal lecture presentation. Across the country, there are now large numbers of professors who routinely use group activities in their large enrollment classes, increasing the level of interaction in their classes, the attention of their students, and the level of satisfaction with the class.

Summary

In summary, there is a large body of research to indicate that judicious use of small groups can significantly enhance the educational experiences of students in chemistry classes and laboratories. While the jury is still out on whether small group learning can significantly affect the achievement of students in chemistry, it is clear that, at the very least, student satisfaction and retention rates are usually increased. Small groups can be used successfully across the whole spectrum in many different courses, from large enrollment introductory classes, to labs and upper-level or even graduate-level courses. While many instructors are reluctant to employ small-group techniques for a variety of reasons, most of these objections can be overcome or are not even valid. The fact is that small groups can be used in so many ways and in so many situations; even the most rigidly constructed course can benefit from interjecting small-group activities into the mix.

Suggested Readings

Cooper, M. M. (1995). Cooperative learning: An approach for large enrollment courses, *Journal of Chemical Education*, 72, 162.
This paper details a number of ways that small groups can be used in a large enrollment chemistry course, including group quizzes and writing to learn assignments. Potential problems and their remedies are discussed.

Felder R. M. and Brent, R. (2001). Effective Strategies for Cooperative Learning. *J. Cooperation & Collaboration in College Teaching,* 10(2), 69–75.
Rich Felder, a chemical engineer, details how cooperative learning can be implemented into a variety of courses and situations

Johnson, D. W., Johnson, R.W., and Smith, K. A.. (1998). *Active learning: Cooperation in the college classroom.* Edina, MN: Interaction Book Company.
The seminal work on cooperative learning in higher education; a resource for almost any situation.

References

Abraham, M. R., and Pavelich, M. J. (2004). *Inquiries into Chemistry, 4th ed.* Waveland Press.

Astin, Alexander W., (1993). *What Matters Most in College.* San Francisco, CA:.Jossey-Bass.

American Chemical Society (1947). *Graduate Training at the DoctoralLevel: Condensed Summary of the Main Report of the Committee on Professional Training* Chemical and Engineering News. 25, 1934–36.

Birk, J., Bauer, R., and Sawyer, D. (2001). *Laboratory Inquiries in Chemistry.* Pacific Grove, CA: Brooks Cole.

Bligh, D.A. (2000). *What's the Use of Lectures?* pp. 44–56 San Francisco CA: Jossey-Bass.

Bodner, G. (1991). I have found you an argument. *Journal of Chemical Education.* 68, 385.

Bodner, G., Klobuchar, M., and Geelan, D. (2001). The Many Forms of Constructivism, *Journal of Chemical Education.* 78, 1107.

Bowen, C.W. (2000). A quantitative literature review of cooperative learning effects on high school and college chemistry achievement. *Journal of Chemical Education.* 77, 116.

Bradley, A. Z., Ulrich, S. M., Jones, M., and Jones, S. M. (2002). Teaching the Sophomore Organic Course without a Lecture. Are You Crazy? *Journal of Chemical Education.* 79, 514.

Corwin A. F., and Wilson, A. R. (1951). Analytical Chemistry and the Cooperative Plan. *Journal of Chemical Education.* 28, 244.

Cooper J., and Robinson, P. (1997). *Small-Group Instruction: An annotated bibliography.* National Institute for Science Education. Available online at http://www.wcer.wisc.edu/nise/. Last accessed October 10, 2003.

Cooper, M. M. (1995).. Cooperative Learning: An Approach for Large Enrollment Courses. *Journal of Chemical Education.* 72, 162.

Cooper, M. M. (2003). *Cooperative Chemistry Laboratories, (2nd Ed.)* McGraw-Hill, NY.

Cueso, J. (1992). Collaborative and Cooperative Learning in Higher Education: A Proposed Taxonomy. *Cooperative Learning and College Teaching. 2(2),* 2–4.

Farrell, J. J., Moog, R. S., and Spencer, J. N. (1999). A Guided-Inquiry General Chemistry Course. *Journal of Chemical Education.* 76, 570.

Felder R. M., and Brent, R. (2001) Effective Strategies for Cooperative Learning. *J. Cooperation and Collaboration in College Teaching. 10*(2), 69–75.

Fiechter, S.B., and Davis, E.A. (1985). Why Groups Fail: A Survey of Student Experiences with Learning Groups. *The Organizational Behavior Teaching Review.* 9(4), 58–73.

Glaser, R E., and Poole, (1999). M. J. Organic Chemistry Online: Building Collaborative Learning Communities through Electronic Communication Tools. *Journal of Chemical Education*. 76, 699.

Greenbowe, T. G., and Burke, K. (2004). Assessing your students understanding of chemistry. In *Survival Handbook for the New Chemistry Instructor*. D. M. Bunce. and C. C. Muzzi (Eds.) Upper Saddle River, NJ: Prentice Hall.

Jackson, P. T., and Walters, J. P. (2000). Role-playing in analytical chemistry: The alumni speak. *Journal of Chemical Education*. 77, 1019.

Johnson, D. W., Johnson, R.W., and Smith, K. A (1998). *Active Learning: Cooperation in the College Classroom*. Interaction Book Company. Edina, MN

Kagan, S. (1992). *Cooperative Learning*. Resources for Teachers Inc. San Juan Capistrano, CA.

Knight G. P., and Bohlmeyer, E. (1992). *Cooperative Learning and achievement: Methods for assessing causal mechanisms.,* In S. Sharon Cooperative Learning Theory and Research (pp. 1–22) NY:Praeger.

Kubli, F. (1983). Piaget's clinical experiments: a critical analysis and study of their implications for science teaching. *European Journal of Science Education*. 5, 123.

Kuhn, D. (1981). The role of self-directed activity in cognitive development. In *New Directions in Piagetian Theory*. D. Siegel. () 353–357. Hilsdale, NJ: Erlbaum..

Lyman, F. (1981). The responsive classroom discussion. In Anderson, A. S. (Ed.) *Mainstreaming Digest*. College Park, MD: University of Maryland College of Education.

MacManaway, L. A. (1970). Teaching Methods in Higher Education: Innovation and Research. *Universities Quarterly*. 24(3), 321.

Mazur, E. (1997). *Peer Instruction*. Upper Saddle River, NJ: Prentice Hall.

McDermott, L.C., and Shaffer, P.S. (2001). *Tutorials in Introductory Physics*. Upper Saddle River, NJ: Prentice Hall.

Meltzer, D. E., and Manivannan, K. (2002). Transforming the Lecture-Hall Environment: The Fully Interactive Physics Lecture. *American Journal of Physics*. 6, 639.

Millis, B.J., and Cottell, P.G., (1998). *Cooperative Learning for Higher Education Faculty,* American Council on Education. Phoenix AZ: Oryx Press.

Mulford, D R., and Robinson, W R. (2002). An Inventory for Alternate Conceptions Among First-Semester General Chemistry Students. *Journal of Chemical Education*. 79, 739.

National Institute for Science Education, web site for collaborative learning at http://www.wcer.wisc.edu/nise/CL1/CL/default.asp. Last accessed August 2003.

National Science Foundation. (1996). *Shaping the Future: New Expectations for Undergraduate Education in Science, Mathematics, Engineering, and Technology*. Washington DC: Report by the advisory committee to the National Science Foundation Directorate for Education and Human Resources.

National Research Council. (1997). *Science Teaching Reconsidered, A Handbook*. Washington DC: National Academy Press.

National Research Council. (2003). *How People Learn*. Washington DC: National Academy Press.

Novak, G., Garvin, A., Christian, W., and Patterson, E. (1999). *Just-in-Time Teaching: Blending Active Learning with Web Technology.* NJ: Prentice Hall.

Nurrenbern, S. C. (1995). *Experiences in Cooperative Learning; A Collection for Chemistry Teachers.* Institute for Chemical Education. University of Wisconsin–Madison

Perret–Clermoat, A. (1980). *Social interaction and cognitive development in children.* London: Academic Press.

Rickey, D., and Stacey, A.M. (2000). The Role of Metacognition in Learning Chemistry. *J. Chem. Educ.* 77, 915–20.

Robinson, W. R., and Nurrenbern, S. C. (1997). Cooperative Learning: A bibliography. *Journal of Chemical Education.* 74, 623.

Scerbo, M. W., Warm, J. S., Dember, W. N., and Grasha, A. F. (1992). The Role of Time and Cuing in a College Lecture. *Contemporary Educational Psychology.* 17(4), 312.

Slavin, R.E. (1996). Research for the Future: Research on Cooperative Learning and Achievement: What we know, what we need to know. *Contemporary Educational Psychology.* 21.

Towns, M.H. (1998). How do I get my students to work together? Getting cooperative learning started.*Journal of Chemical Education.* 75, 69.

Towns, M. H. (2001). Kolb for Chemists: David A. Kolb and Experiential Learning Theory. *Journal of Chemical Education.* 78, 1107.

Towns, M., Sauder, D., Whisnant, D., and Zielinski, T. J. (2001). Physical Chemistry Online. *Journal of Chemical Education.* 7,8, 414.

Vokos, S., Shaffer, P. S., Ambrose, B. S., and McDermott, L.C. (2000). Student Understanding of the Wave Nature of Matter: Diffraction and Interference of Particles. *American Journal of Physics.* 68, S47.

Walters, J. P. (1991). Roleplaying Analytical Chemistry Laboratories: Parts I, II, and III. *Analytical Chemistry.* 63, 977A, 1077A, and 1179A .

Wheatley, G. H. (1991). Constructivist perspectives on science and mathematics learning. *Science Education.* 75, 9.

Williamson, V. M., and Rowe, M. W. (2002). *Journal of Chemical Education.* 7, 9, 1131.

Wilson, G. (2003). *Exploring the Molecular Vision.* Conference, Washington DC.

Wink, D. J., Fetzer–Gislason, S., and Ellefson–Kuehn, J. (2000). *Working with Chemistry.* New York: W. H. Freeman.

Wolfskill, T., and Hanson, D. (2001). LUCID: A New Model for Computer-Assisted Learning. *Journal of Chemical Education.* 78, 1417.

Wright, J. C. (1996). Authentic learning environment in analytical chemistry using cooperative methods and open-ended laboratories in large-lecture courses. *Journal of Chemical Education.* 73, 827.

11

Using Concept Maps to Figure Out What Your Students are Really Learning

Mary B. Nakhleh
Department of Chemistry
Purdue University

Yilmaz Saglam
Purdue University

Abstract

Students often have trouble understanding that the different concepts taught in chemistry really do have an underlying connection. Concept maps are one way to help students make meaningful and complex connections between the various concepts that make up the topics of chemistry. This chapter discusses how to teach students to make concept maps and how these maps can be used in both instruction and assessment to help students develop a more complete and robust understanding of chemistry.

Biographies

Mary Nakhleh received her B.S. in chemistry at the University of Georgia in 1961 and started her professional career as an industrial chemist in Falls Church, Virginia. Later, she became a middle school teacher and spent three years teaching life science, earth science, and math to seventh and eighth graders. She then became a high school teacher of chemistry, physical science, and computer science for 12 years in Maryland. Along the way, she became interested in the problems of chemistry education and received her M.A. (1985) and her Ph.D. in science education (1990) from the University of Maryland, College Park. Since 1990, she has been a jointly appointed faculty member at Purdue University in West Lafayette. She is currently Associate Professor of Chemistry and Curriculum and Instruction. She often uses concept maps in her research to investigate how students use technology as a learning tool.

Yilmaz Saglam received his bachelor's degree in chemical education from Selcuk University in Konya, Turkey in 1997. He taught science to middle school students for almost two years before coming to the United States to study. He received his master's degree in science education from the University of Missouri in 2001, and he is currently pursuing his Ph.D. in science education at Purdue University. He is interested in using concept maps as a learning tool.

Ideas about Learning

As educators, whether professors or teachers or teaching assistants, most of us have often asked ourselves a very disturbing question: *How do I know what my students are learning about this topic?* Oh, we notice that they seem to be paying attention, even taking notes and asking questions, and they perform more or less well on exams. Yet, the question persists: *I wonder how much they're really learning?* This chapter will discuss how you might use concept maps in your course to begin to address this question. Don't be surprised if the answers you uncover are sometimes heartening and sometimes surprising!

Concept maps are a way of creating a visual network of the set of interrelationships of events, objects, and ideas that describe a person's understanding of a topic, such as acids. Here, we use Novak and Gowin's (1984, p. 4) definition of a concept as "a regularity in events or objects designated by some label." In other words, we know how to look at an object and classify it as an "acid" or "base" by using a set of ideas about what an "acid" or "base" should be. Notice that, in this definition, a concept involves a set of ideas, which are themselves concepts. These maps were introduced to the science education research community by Novak and Gowin (1984); they come out of the constructivist tradition in educational research. Nakhleh (1992) points out that, in the constructivist tradition, learners are viewed as thinking individuals. Further, learning in a course is viewed as much more than the simple transmission of information. Learners are considered to be thinking entities who actively construct their understanding from their prior experiences and the information that they receive in instruction. In this view, information is not knowledge, but this information can be processed and integrated into students' current knowledge framework, thereby becoming part of that framework. Therefore, a concept map can potentially be a useful tool to encourage our students to actively engage in learning and to assist us in assessing the depth and breadth of that learning.

Creating a Concept Map

Concept maps are composed of two main elements: concept nodes and relationship lines (Nakhleh, 1992). The nodes are often what an older tradition would call "vocabulary terms," and the relationship lines contain a short descriptive phrase or sentence that explains how two nodes are related. The small concept map in Figure 1 illustrates how a few concepts and relationships can create a simple but informative concept map. Figure 1 clearly shows how the person who made the map related the five concept nodes to each other, although we do not know which node was the central concept in his or her thinking.

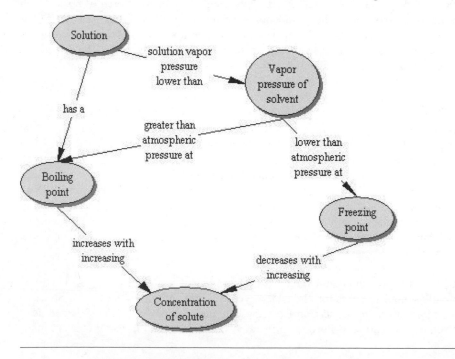

Figure 1. A simple concept map about solutions.

However, most concept maps are more elaborate than this one, and it is usually helpful to indicate the central concept of your topic by a special symbol. Often, identifying this central concept is the hardest part of drawing a concept map. Sometimes you have to experiment with different central concepts until you have a satisfactory map. Then you connect the central concept to other nodes containing concepts that relate to and/or describe your central concept. These nodes are then connected by relationship lines, and short phrases describing the relationship are written on the lines. We use arrowheads to indicate direction on these relationship lines. The arrowheads can relate two nodes by ← or → or even ←→. You continue to construct your map by spiraling out from the central concept. In other words, every concept node on the map should have a traceable relationship line connection back to the central concept. Concept nodes can have many directions for their relationship lines, but there should be at least one way to trace each concept back to the central concept of the map. Then you continue to interconnect the concepts with relationship lines until you have displayed your knowledge of the topic as completely as you can.

If you have some familiarity with concept maps, you might have noticed that we did not emphasize any hierarchy of concepts in these maps. That is because we do not believe that chemistry is by nature a very hierarchical discipline, so we choose to use a concept network more than a hierarchical map. Early concept maps in science education were constructed in the field of biology, which in some cases lends itself better to hierarchy than chemistry. Chemistry concepts seem to us to integrate across topics rather than descend in a vertical hierarchy. In earlier work, Nakhleh (1994) had tried to use hierarchy in analyzing secondary students' concept maps of acids, bases, and pH. However, she found that students did not draw hierarchical maps and that the maps were more useful when hierarchy was ignored. However, an appropriate concept map does have every concept flowing from the central concept. For example, all concepts shown in Figure 2 can be traced back to the central concept of "solution."

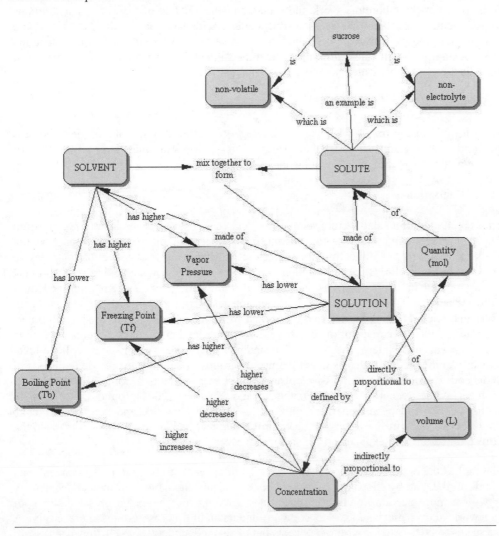

Figure 2. A more complex concept map about solutions.

One of the most important ideas about concept maps is the notion that whenever we retrieve any one of the concepts in this network, *we have the potential to access all of the concepts stored in that network*. Therefore, students who have constructed maps similar to this one have a visual display of how these concepts relate to each other and to the central idea of solution.

Practical Ways to Use Concept Maps

A large body of literature describes the many ways in which concept maps have been tried to enhance student learning. For example, Regis, Albertazzi, and Roletto (1996) explored the use of concept maps with secondary chemistry students and reported that the maps helped the students become aware of and critique their own understanding. Gayle Nicoll (Nicoll, Francisco, and Nakhleh, 2001) used concept maps to assess college students' understanding of the nature of chemical bonding. Markow and Lonning (1998) investigated the usefulness of concept maps in college chemistry laboratories. They used a control group of students who wrote essays explaining the chemical concepts in a lab and an experimental group that drew pre- and post-lab concept maps. They found that students responded very positively to the concept maps, reporting that the maps helped them improve their conceptual understanding of the chemistry in the labs. However, Markow and Lonning also note that they found no significant differences between treatment groups in their achievement scores.

However, college-level courses have specific constraints. Some institutions have lecture sections that are large (Nakhleh's 400+ lecture, for example); others will accommodate at most 20–30 students. In some institutions, the professor is highly involved in the laboratory work and seeing students outside of class; other institutions depend more heavily on teaching assistants as the front-line contact with students in the laboratory and recitations. Here, we will explore the ways in which concept maps can be used to provide both the student and the professor or teaching assistant with valuable information about the students' understanding of chemistry. Of course, you will need to adjust the ideas presented here to your own situation.

Concept Maps as an Instructional Tool

Concept maps can play a useful variety of roles in instruction (Regis, Albertazzi, and Roletto, 1996; Novak, 1991; Edmondson and Smith, 1998; Romance and Vitale, 1999). This section examines some of the uses of concept maps in the lecture/recitation setting and then in the laboratory.

Groupwork in Lecture or Recitation

In smaller settings, such as recitation or a small lecture section, we believe that students can benefit greatly from working in small groups of three to five students to construct a concept map on a specific topic.Specific university-level chemistry examples from the literature are scarce, but work done in lower-level biology classes is very suggestive. For example, Guastello, Beasley, and Sinatra (2000) found that low-achieving seventh grade students who actively participated in constructing a concept map of the circulatory system improved their performance on standardized and teacher-constructed comprehension tests by six standard deviations. Also, Osmundson, Chung, Herl, and Klein (1999) reported that fourth and fifth grade students who used a computer-based mapping tool to create collaborative maps of the digestive, respiratory, and circulatory systems developed more scientific, principle-based, and interconnected understandings of these systems. In college-level chemistry, Nakhleh, Lowery, and Mitchell (1996) studied the use of groups in solving open-ended conceptual and algorithmic problems and found that the groupwork apparently did help students become more proficient conceptual problem solvers. More recently, Nicoll, Francisco, and Nakhleh (2001) found that freshman chemistry students who used concept maps in homework, quizzes, recitation groupwork, and exams could make more complex connections between chemistry topics than a control group that had not used concept maps in the course. Chapter 11 of this publication also expands on the use of groupwork in chemistry instruction.

In groupwork, the professor has the option of giving each group a list of concepts to be used in the concept map or letting the student freeform the map by choosing their own terms. Students choosing their own concepts may provide the instructor with a more accurate picture of their understanding, but having a preselected set of concepts with which to work allows the instructor to compare understanding across groups more easily. A reasonable compromise seems to be to provide a set of concepts but to also tell students that (1) they can discard concepts they don't think fit in their map and (2) they can add new concepts if they think that the new concepts are needed.

As a practical point, a useful way to work on concept maps is to hand out Post-It notes, markers, and large sheets of paper. Students should write their concepts on the Post-It notes and move them around on the paper as they debate and argue over the placement of the concepts and what to write on the relationship lines. When they reach agreement, they can tape down the notes on the paper and then draw in the relationship lines.

As the students work on their maps, one of the real benefits of this activity is that they have an opportunity to talk about the concepts, argue out their ideas about how the concepts relate to one another, and defend their ideas before the group. Jay Lemke (1990) calls this "talking science" and argues that this mode of debate helps students clarify their ideas. While the students are working on the maps, the instructor should walk around and observe the maps. Are there gaps or weaknesses in the network of concepts being created? Gaps in understanding can be shown by the absence of key concepts on the maps or by very few or weak links to the concepts. Weaknesses in understanding can be seen by observing the degree to which the concepts are crosslinked and the usefulness of the connecting links. For example, Figure 3 clearly shows that the person making the map did not demonstrate any understanding of how "solution" relates to the other concepts in the map.

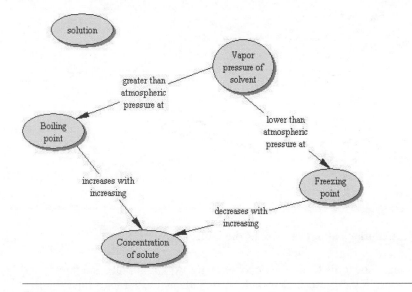

Figure 3. A concept map indicating partial understanding.

Kinchin, Hay, and Adams (2000) also discuss how the overall structure of a concept map can help instructors and students analyze their understanding. For example, a very linear map with no crosslinking may indicate a linear understanding, i.e., a concept is only linked to those immediately before and after it. A very crosslinked map might indicate a deep understanding of the major concepts in a topic.

Also, the instructor should look over the maps to see if any misconceptions have been represented. Misconceptions may be revealed by inappropriate links to other concepts. For example, the map in Figure 4 contains two misconceptions: (1) the size of molecules is influenced by heat, and (2) the size of a molecule can change depending on the physical state. The concept map provides a convenient means of discussing these misconceptions with the students. In fact, if the map has been created by a group, so much the better. Then you can discuss the misconceptions without having to single out individuals.

You can also estimate how firmly entrenched the misconceptions are. In the small example map in Figure 4, the size of molecules is unfortunately connected to two different concepts: heat and states of matter. Therefore, these misconceptions about size will probably be more difficult to change because the change involves breaking and rearranging two links rather than one.

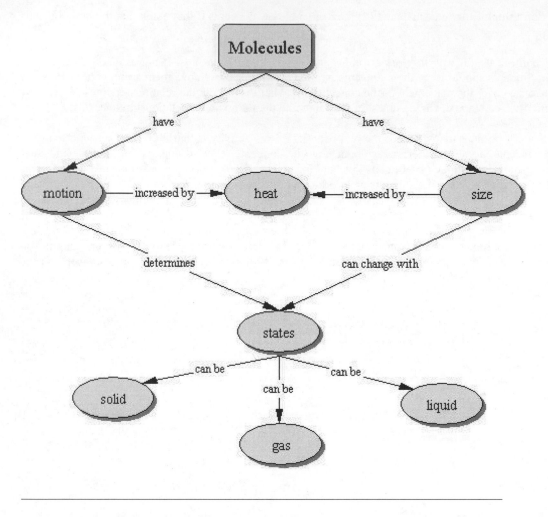

Figure 4. A concept map showing misconceptions about the size of the molecules.

Students can learn to use their maps as diagnostic tools to aid their own learning. At a minimum, students could use the maps to identify areas of weaknesses in their own understanding. If students have a hard time figuring out links between concepts or are forced to use very general connective phrases, then they should realize that this process identifies areas of weakness in their own understanding. White and Gunstone (1992) also provide a good discussion of how concept maps can be used by either students or teachers in analyzing a learner's understanding.

Concept maps in the laboratory

Concept maps can also be used to help students understand the concepts underlying laboratory experiments (Markow and Lonning, 1998). For example, the pre-lab questions could ask students to identify the chemistry concepts that are important to the laboratory and then to arrange those concepts into a concept map that would relate the chemistry concepts to the laboratory procedures.

The concept map of chromatography in Figure 5 shows one way that this might be done. By counting the number of incoming lines, the map shows that there are five key concepts important to this lab: mobile phase, stationary phase, retention times, Rf values, and standards. Therefore, students could focus on understanding these concepts and discussing how important these concepts are to the lab. Then in the post-lab questions, students could have the opportunity to modify their concept map as part of their laboratory report. In this way, students could use the concept map to reflect on the meaning of the experiment and on how to relate the laboratory procedures and observations to the concepts learned in lecture.

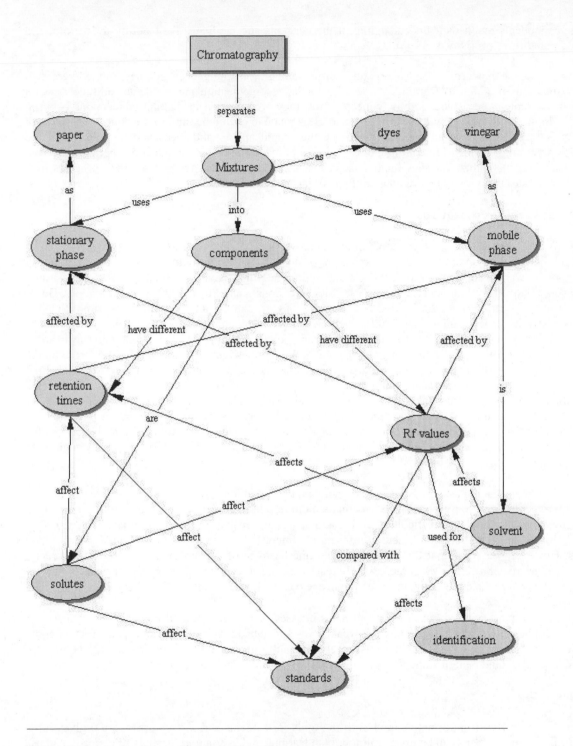

Figure 5. A concept map of a laboratory on chromatography.

Concept Maps as a Learning Tool

Concept maps can also help students create a meaningful understanding of science rather than simply memorizing scientific "facts." Novak (1995, 1998) argues that meaningful learning occurs when the learner makes connections between newly learned understanding and prior knowledge. However, memorization occurs when a student learns new information without relating this information to prior understanding. A conceptual network helps students retain their understanding of the concepts and facilitates their ability to interpret and understand related concepts (Novak, 1998).

Concept maps can be powerful tools in creating this meaningful learning because they can address two important aspects of meaningful learning: students' knowledge organization and students' attitudes toward learning. For example, when students draw concept maps, they are involved in making connections between concepts and in thinking about how these concepts and ideas relate to each other and to the conceptual network as a whole. Students also have an opportunity to see their valid and invalid ideas and to think about the relationships between those ideas. In addition, drawing concept maps and making connections between relevant concepts make students realize that they have some relevant knowledge, and they can see their progress over time, which motivates them toward more meaningful learning (Novak, 1998).

Concept Maps as an Assessment Tool

Concept maps can be used as assessment tools in two main ways: as formal (quantitative) assessments and as informal (qualitative) assessments. Many researchers have investigated using them as a formal assessment in which points are awarded for appropriate concepts and appropriate links (Liu and Hinchey, 1996; Ruiz-Primo and Shavelson, 1996; Rice, Ryan and Samson, 1998; McClure, Sonak, and Suen, 1999; Robinson, 1999; Bolte, 1999; Shavelson and Ruiz-Primo, 2000; Stoddart, Abrams, Gasper, and Canaday, 2000; Ruiz-Primo, Schultz, Li, and Shavelson, 2001; Rye and Rubba, 2002). There have been many attempts to devise scoring schemes for concept maps, and the one presented here was adapted from Novak and Gowin (1984). One point is awarded for every acceptable relationship, including examples. Crosslinking between concepts in different areas on the map are an indication of integrating knowledge, which indicates a substantial growth in understanding. Crosslinks are scored at ten points each. Using this scoring system, the concept map in Figure 4 would receive 10 points for one acceptable crosslink, and six points for six acceptable relationships (links). The score for acceptable links is 16 points. However, you should also note that one crosslink and one relationship (link) are unacceptable (total of 11 points). Therefore, you could choose to deduct 11 points from the acceptable relationship score to give a total score of only 5 points. So you can see that concept map scoring is fraught with many decisions.

Notice that if a line is counted as a crosslink, it does not count as a relationship. The total raw score can then be converted to percentages if you wish. Notice also that we did not discuss how to award points for hierarchical organization because hierarchy is difficult to define and score, and it may not be appropriate for chemistry. Novak and Gowin include a complete discussion if you are interested in scoring the hierarchical organization ofconcepts. However, we argue that this is probably not the best use of concept maps, which are inherently qualitative and idiosyncratic. For example, two students can construct maps that on the surface look very different, yet closer analysis could show that the knowledge content is equivalent.

We believe that the most powerful use of concept maps in assessment is in informal contexts. In other words, both students could use concept maps to identify weaknesses and gaps in their understanding; instructors could also use the maps to spot potential or actual misconceptions built into the students' understanding. Hopefully, this type of feedback could benefit students and instructors before any formal exams occur.

Conclusions

Concept maps can be useful, informative, and even fun tools for students and instructor. Figure 6 gives some examples of how concept maps can be used to further both learning and instruction. A study of concept map use in college-level chemistry (Nicoll, Francisco, and Nakhleh, 2001) indicated that students who used concept mapping in the course fared significantly better than students who did not use concept mapping. However, there are a couple of important ideas to consider before implementing concept maps in your course. The most important idea is *Don't implement every idea all the time.* You want to aim for evolution, not revolution. The second idea is that groupwork is probably the most readily accessible type of concept mapping activity to you and to your students. Also working in groups tends to stimulate communication and improve attitudes in a course. Finally, please remember that you have to train students to do these maps; they don't come immediately or naturally to everyone. Again, groupwork is a good way to accomplish this training because there are many avenues for feedback.

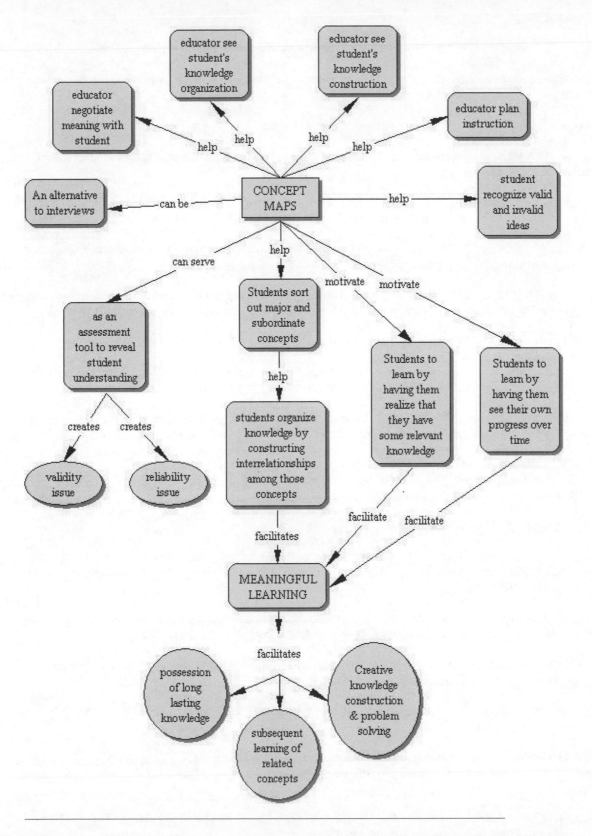

Figure 6. Possible roles of concept maps in learning and instruction (adapted from Novak, 1998).

Suggested Reading

Novak, J. D. & Gowin, D. (1984). *Learning How to Learn*. New York, Cambridge University Press.

A monograph on what concepts maps are, how to get students to start using concept maps, and how to evaluate concepts maps. This is the book that started it all.

References

Bolte, L. A. (1999). Using concept maps and interpretive essays for assessment in mathematics. *School Science and Mathematics.* 99: 19–30.

Edmondson, K. M., and Smith, D. F. (1998). Concept mapping to facilitate veterinary students' understanding of fluid and electrolyte disorders. *Teaching and Learning in Medicine.* 10(1): 21–33.

Guastello, E. F., Beasley, T. M., and Sinatra, R. C. (2000). Concept mapping effects on science content comprehension of low-achieving inner-city seventh graders. *Remedial and Special Education.* 21(6): 356–365.

Kinchin, I. M., Hay, D. B., and Adams, A. (2000). How a qualitative approach to concept map analysis can be used to aid learning by illustrating patterns of conceptual development. *Educational Research.* 42(1): 43–57.

Lemke, J. L. (1990). *Talking Science: Language, Learning, and Values*. Norwood, NJ: Ablex.

Liu, X., and Hinchey, M. (1996). The internal consistency of a concept mapping scoring scheme and its effect on prediction validity. *International Journal of Science Education.* 18: 921–937.

Markow, P. G., and Lonning, R. A. (1998). Usefulness of concept maps in college chemistry laboratories: students' perceptions and effects on achievement. *Journal of Research in Science Teaching.* 35: 1015–29.

McClure, J. R., Sonak, B., and Suen, H. K. (1999). Concept map assessment of classroom learning: reliability, validity, and logistical practicality. *Journal of Research in Science Teaching.* 36: 475–492.

Nakhleh, M. B. (1992). Conceptualization and assessment of laboratory experiences: The use of Vee diagrams and concept maps. *The Hoosier Science Teacher.* 8(1): 5–12.

Nakhleh, M. B., and Krajick, J. S. (1994). Influence of levels of information as presented by different technologies on students' understanding of acid, base, and pH concepts. *Journal of Research in Science Teaching.* 31(10): 1077–1096.

Nakhleh, M. B., Lowrey, K. A, and Mitchell, R. C. (1996). Narrowing the gap between concepts and algorithms in freshman chemistry. *Journal of Chemical Education.* 73(8): 758–762.

Nicoll, G., Francisco, J., and Nakhleh, M. (2001). An investigation of the value of using concept maps in general chemistry. *Journal of Chemical Education.* 78: 1111–17.

Novak, J. D. (1991). Clarify with concept maps: A tool for students and teachers alike. *The Science Teacher.* 58(7): 44–49.

Novak, J. D. (1995). Concept mapping: a strategy for organizing knowledge. In S. M. Glynn and R. Duit (Eds.) *Learning Science in the Schools: Research Reforming Practice.* 229–245. Mahwah, NJ: Lawrence Erlbaum Associates.

Novak, J. D. (1998). *Learning, Creating, and Using Knowledge: Concept Maps as Facilitative Tools in Schools and Corporations*. London: Lawrence Erlbaum Associates.

Novak, J. D., and Gowin, D. (1984). *Learning How to Learn.* New York: Cambridge University Press.

Osmundson, E., Chung, G.K.W.K., Herl, H.E., and Klein, D.C.D. (1999). Knowledge mapping in the classroom: A tool for examining the development of students' conceptual understandings (Tech. Rep. No. 507). Los Angeles: University of California, Los Angeles, Center for the Study of Evaluation.

Regis, A. , Albertazzi, P. G., and Roletto, E. (1996). Concept maps in chemistry education. *Journal of Chemical Education.* 73: 1084–88.

Rice, D. C., Ryan, J. M., and Samson, S. M. (1998). Using concept maps to assess student learning in the science classroom: Must different methods compete? *Journal of Research in Science Teaching.* 35(10): 1103–1127.

Robinson, W. R. (1999). A view from the science education research literature: concept map assessment of classroom learning. *Journal of Chemical Education.* 76: 1179.

Romance, N. R., and Vitale, M. R. (1999). Concept mapping as a tool for learning: broadening the framework for student-centered instruction. *College Teaching.* 47(2): 74–79.

Ruiz–Primo, M. A., Schultz, S. E., Li, M., and Shavelson, R. J. (2001). Comparison of the reliability and validity of scores from two concept-mapping techniques. *Journal of Research in Science Teaching.* 38: 260–278.

Ruiz–Primo, M. A., and Shavelson, R. J. (1996). Problems and issues in the use of concept maps in science assessment. *Journal of Research in Science Teaching.* 33(6): 569–600.

Rye, J. A., and Rubba, P. A. (2002). Scoring concept maps: An expert map-based scheme weighted for relationships. *School Science and Mathematics.* 102(1): 33–44.

Shavelson, R. J., and Ruiz–Primo, M. A. (2000). On the psychometrics of assessing science understanding. In J. Mintzes, J. Wandersee, and J. Novak (Eds.) *Assessing science understanding.* (pp. 303–341). San Diego, CA: Academic Press.

Stoddart, T. Abrams, R., Gasper, E., and Canaday, D. (2000). Concept maps as assessment in science inquiry learning-a report of methodology. *International Journal of Science Education.* 22: 1221–1246.

White, R., and Gunstone, R. (1992). *Probing Understanding.* London: Falmer Press.

12

Introduction to the Science Writing Heuristic

Thomas J. Greenbowe
Department of Chemistry
Iowa State University of Science and Technology

Brian Hand
Department of Curriculum and Instruction
Iowa State University of Science & Technology

Abstract

This chapter presents an overview of the theory of writing to learn science and the use of the Science Writing Heuristic as a tool for structuring chemistry laboratory activities and the chemistry laboratory report. A heuristic is a general overarching theme that governs how individuals approach or view an event. The Science Writing Heuristic contains both teacher and student templates for structuring appropriate guided-inquiry activities through the use of writing and collaborative peer discussion. The Science Writing Heuristic is designed to help students make connections among research questions, experimental procedures, data, knowledge claims, evidence, and science concepts. When the Science Writing Heuristic is implemented by effective instructors and embraced by the majority of students in a class, average student performance on chemistry examinations and laboratory practical tasks improves compared to a class of students who have an instructor who follows a traditional or verification approach to chemistry laboratory activities and uses a standard laboratory report format.

Biographies

Thomas Greenbowe is currently professor of chemistry and coordinator of general chemistry at Iowa State University. He teaches general chemistry to large sections of science and engineering students. Over the past years, he has taught more than 20,000 students. He tries to make his lecture presentations as interactive as possible, employing group work, multimedia, demonstrations, and humor. He works with teaching assistants to help them implement active learning in their recitation and laboratory sections. Before coming to Iowa, Tom was a faculty member at the University of Massachusetts–Dartmouth, the chemistry lecture demonstrator at Purdue University, and a high school physics and chemistry teacher. His Ph.D. is from Purdue University. Tom has served on several ACS General Chemistry Exam committees and is currently the general chair of the 18th Biennial Conference on Chemical Education. Tom is delighted to learn effective teaching and learning techniques from colleagues across the country. He shares what works in his classroom by speaking at ACS meetings and college seminars. He has published papers about chemical education issues and research in the *Journal of Chemical Education, Journal of Research in Science Teaching, Journal of College Science Teaching,* and the *International Journal of Science Education.* He has served as a principal investigator for several curriculum and development projects funded by the National Science Foundation and by the U. S. Department of Education. Making changes in the way we teach chemistry laboratories continues to be a challenge, presenting a moving target.

Brian Hand is a professor of science education and Director of the Center for Excellence in Science and Mathematics Education in the Department of Curriculum and Instruction at Iowa State University. His research interests are in the use of writing to learn strategies within science classrooms, the Science Writing Heuristic, and improving teaching and learning in science. Prior to joining the faculty at Iowa State University, he taught high chemistry, physics, and junior science for 11 years in three different states of Australia. He began his tertiary career at LaTrobe University, Bendigo, in Australia. He has published papers in the *Journal of Chemical Education, Journal of Research in Science Teaching, Science Education, Journal of College Science Teaching,* and the *International Journal of Science Education.* He currently serves on the editorial board of the *Journal of Research in Science Teaching.* He has been involved as principal investigator on a number of National Science Foundation and by the U. S. Department of Education grants.

Introduction

Introductory chemistry courses for college students tend to emphasize numerical problem solving during lecture presentations and on examinations. Instructors usually give two reasons for doing so. First, instructors believe that by solving problems, students learn chemistry. Second, it is easier and quicker to grade an examination problem when there is one correct answer. Instructors do not have to justify why an answer is wrong. There is little or no negotiation between the student and the instructor as to the meaning of the student's answer. As a chemistry instructor, you know that if you ask students to write an explanation to accompany their answer, you are in for a long night of grading and you will learn just how much your students don't know about the topic addressed by the problem. This situation occurs because most instructors do not assign homework problems or quiz problems that require written explanations. If students are not given the opportunity to practice writing explanations prior to examinations that have some written explanations as part of a problem, poor performance is the rule. Is it worth the time and effort to require your students to provide written explanations for chemistry homework, examinations, and laboratory reports? Can students actually learn chemistry, in part, by writing explanations to chemistry events? We believe the answer to these two questions is yes, if you and your students implement the appropriate techniques of guided-inquiry, group work, and the Science Writing Heuristic. An excellent place to start is in the chemistry teaching laboratory.

Most chemistry instructors assume that if students do a chemistry laboratory activity they will learn something, that is, by following the "directions" given in the laboratory manual, students will learn. Over the past 20 years, science education researchers have investigated what students' gain from science laboratory experiences (Hofstein and Lunetta 1982; Nakhleh, Polles, Malina 2002). One consistent finding of educational research studies is that if laboratory experiments are used with the traditional laboratory notebook format and with traditional verification activities, students may learn laboratory techniques, but they learn little else (Lazarowitz and Tamir 1994). If traditional laboratory activities are used to demonstrate theory or concepts, students do not engage in these activities in ways that are reflective of the scientific dialogue and critical debate that scientists use to arrive at the demonstrated theory or concept. Gunstone (1990) stated that "for practical (laboratory) work to have any serious effect on student theory construction and linking concepts in different ways, the students need to spend more time interacting with ideas and less time interacting with apparatus."

A problem with traditional laboratory experiments and with traditional laboratory reports is the fact that students often will blindly follow the directions given in the laboratory manual, view the data collected during a laboratory experiment as artificial, plug in numbers to a formula to generate an answer or identify an unknown, and then compare their answer with "the literature value." When the answer generated from data they collect has a large percent error or the identity of their unknown is incorrect, their discussion section of the laboratory report may blame poor laboratory equipment, human error, or chance.

We ask you to consider why average student performance on a specific topic on your examination is poor. Performance will be poor even when the examination problems closely match what you presented in lecture, what you assigned for reading in the textbook, what was on the assigned homework, what the students did in the laboratory, and the questions that were on a quiz. If your goal as a chemistry instructor is to provide opportunities for your students to learn chemistry, then you need to think carefully about the purpose of the laboratory component of your course. This chapter will provide an introduction to the theory of why writing, in part, helps students learn chemistry. Using the Science Writing Heuristic as the basis for your guided-inquiry laboratory experiments, along with group work, is the key to helping students increase their conceptual understanding of, and improve their attitude toward, chemistry.

Models of Writing As a Method for Learning Science

Classic cognitive model. There are a number of important concepts that need to be dealt with when discussing the concept of using writing as a learning tool. Writing is not speech written down, but rather, "writing represents a unique mode of learning—not merely valuable, not merely special, but unique" (Emig, 1977, p. 122). Writing can be viewed as either a knowledge-telling or a knowledge-transforming process (Bereiter and Scardamalia ,1987). In essence, the knowledge-telling model is viewed as a recall process (a regurgitation of facts of sorts), and does not involve the transforming of existing knowledge. The reason no knowledge transformation occurs is because "contents are retrieved on the basis of their organization in memory and the discourse schemas stored by the writer and are then translated directly into text" (Tynjala, Mason, and Lonka, 2001). Thus, any writing that merely requires recall, note taking, or answering study questions is not going to be as valuable in promoting learning because little cognitive demand is placed on the student. Such writing is about demonstration of the knowledge that already exists within the learner. In a landmark study, Langer and Applebee (1987) found that teachers of science, as opposed to other subjects, tended to use these writing activities that focused on factual information rather than the underlying concepts.

When comparing experts and novices on writing tasks, experts develop an elaborate set of goals for their written passage and then generate ideas in order to satisfy these goals. Novices simply recall sentences or fragments and then attempt to translate or re-write these sentences into a written passage (Flower and Hayes, 1980). This is called the classic cognitive model of writing.

For example, most writing assignments or written explanations on examinations, if given, are returned to the instructor for grading. Students use the language that they think the instructor wants to see, that is, science terms with little of the student's own explanations. There is a need to ensure the rhetorical demands required in a textbook format are met, that is, writing for a graded assignment is not free writing. Students are required to meet the need to limit the amount of text written, to structure their writing in a manner required by the instructor (which mimics the textbook format), and to play it safe by not revealing any personal misconceptions. Students will often submit written passages to be graded even if they do not believe or understand what they have written.

Knowledge constitution model. Galbraith (1999) introduced a knowledge constitution model to describe understandings that develop as a result of writing. He argues that new or novel ideas, relative to an individual, are generated by the sentence production process involved when an individual needs to communicate in writing to satisfy a goal. This sentence production routine operates on a sub-propositional level of representation. Part of his model involves a cyclical process involving a content knowledge area, a linguistic network, sentence propositions or clauses, a content task, an activation area, and a constraint. In this model, "ideas are not stored explicitly, but emerge in context as transient stable states of the network as a whole. Furthermore, "novel" ideas emerge automatically whenever the network responds to a novel set of input constraints" (Galbraith, 1999, p. 145.). When a writer wants to convey a message by writing, the message provides the input to a linguistic network, which is responsible for generating sentences or clauses. Hence, in this model, there are four overlapping processes: (1) activating content knowledge, (2) resolving the content into a message, (3) resolving the message into a sentence, and (4) providing feedback. Individuals evaluate their writing and change what they write when they decide that what is written does not match what they believe. Also, individuals receive a critique of their writing from other people. Acting on feedback causes the cycle to repeat. In this model, content can be produced in two ways: it can be recalled from memory or it can be generated in the course of writing.

In order to test this model, Galbraith (1996) did an experiment involving three groups of students. Group A students were told to first generate a list of ideas, second to write a sentence that summed up their response to the topic, then to write an essay on the topic. Group B students were asked to write an essay without worrying about how organized the essay might be. Group C students were asked to first write an outline of the topic, then to write the essay. Group A, who were high self-monitors, outperformed high self-monitored students in the other two groups. This experiment provides some rationale to encourage students to write an outline of what they will be writing about and to summarize the main outcome or idea before writing an essay on that topic.

Knowledge transforming model. On the other hand, a writing assignment can be structured to encourage the development of conceptual understanding of a topic, rather than an exposition of facts. In order to do this, instructors should allow students to initially verbalize and to write about their personal understanding of science without penalty or ridicule.   If students initially say something or write something that is not scientifically correct, the learners should work on an activity that is designed to show them that what they are thinking does not match what actually happens.

A student can be asked to write an explanation of a topic for his or her peers or to students in a lower grade level. Both their peers and the instructor can be involved in the critique and evaluation of the writing assignment. After their written work is critiqued, students are given the opportunity to discuss their written work with their peers and to revise their writing. Students can then work in small groups on an assignment that involves a computer simulation, a laboratory experiment, a hands-on activity, or a written tutorial that encourages problem solving and a verbal and written explanations of events. This process is called a knowledge-transforming model.

Bereiter and Scardamalia (1987) attribute the discovery of insight while writing to active problem solving. They argue that when an individual composes an original written passage, he or she actively deals with conceptual understanding by linking new knowledge to existing concepts. This results in the transformation of old knowledge to new knowledge. In particular, Elbow (1973) argued that the discovery of ideas, insights, or "this doesn't make sense" occurs when students are encouraged to write a first draft without worrying about grammar, organization, or correctness of the content knowledge. Understanding occurs when an individual uses problem-solving skills, rhetorical skills, and content knowledge to rewrite and refine his or her written work.

When an activity or writing assignment is structured in this manner, there are a number of factors with which the students must deal. First, the students need to ensure that written paragraphs can be understood by their peers; that is, there is a need to expand on the terms to be discussed for the audience chosen. The writers are free to use analogies and personality characteristics of objects that most likely are not scientifically correct. Second, the written passages will need to be interesting and relevant to the reader. Third, the writers need to believe and understand what they have written. Fourth, the writers need to follow the rules for grammar, syntax, and logical construction of paragraphs. By thinking about what they know about a topic, what they don't know about a topic, what makes sense to them, what does not make sense to them, and how to write it using their rhetorical skills, the writers will "construct" or develop a deeper level of knowledge of the topic (Hand, 2004). Constructing a piece of text requires the learners to move backwards and forward between these areas in a dynamic fashion. Work completed in one area shapes the work to be completed in the other area (Keys, 1999). This process helps students move past recall and engage cognitively with the subject matter knowledge about which they are writing (Hand, 2004).

Another important consideration highlighted by Keys (1999) is the constraint of the nature of the discipline when using writing as a learning tool. Science is a discipline that is based on particular patterns of argument incorporating the need for claims and evidence. As such, any writing used within the field of science needs to ensure that there is adherence to what constitutes the essence of the discipline. The Science Writing Heuristic is designed to channel what occurs during structured writing by using aspects of the knowledge-constituting model and the knowledge-transforming model of writing (Galbraith, 1999). By using these two models, instructors can enable students to develop a better understanding of the content knowledge about which they are writing.

The Science Writing Heuristic

The Science Writing Heuristic (SWH) can be understood as both an alternative format students use for their laboratory reports and a teaching technique used by the instructor (faculty or teaching assistant) to help format the flow of activities associated with an experiment (Poock, Burke, Greenbowe, and Hand, 2003; Burke, Poock, Greenbowe, and Hand, 2004). A heuristic can be a guide or a method used to help individuals or people to discover or reveal something.

When using the SWH, the role of the instructor changes. Instead of the laboratory instructor telling the student what to do, acting as the source of all of the correct information, and passing judgment on the percent error of a student's product, he or she serves as a facilitator and helps guide students to design an experiment to answer the student's questions and to develop conceptual understanding. For students using the SWH, the biggest change is in writing the laboratory notebook. Instead of responding to the five traditional sections (purpose, methods, observations, results and conclusions), students are expected to respond to prompts eliciting their questions (related to the activity, knowledge claims, evidence, description of data and observations, and methods), and to reflect on changes to their own thinking. Figure 1 provides an overview of the student template and the teacher template for the SWH.

The Science Writing Heuristic, Part I A template for teacher-designed activities to promote laboratory understanding.	The Science Writing Heuristic, Part II A template for the student.
1. Exploration of pre-instruction understanding through individual or group concept mapping or working through a computer simulation. 2. Pre-laboratory activities, including informal writing, making observations, brainstorming, and posing questions. 3. Participation in laboratory activity. 4. Negotiation Phase I—writing personal meanings for laboratory activity (for example, writing journals). 5. Negotiation Phase II—sharing and comparing data interpretations in small groups (for example, making a graph based on data contributed by all students in the class). 6. Negotiation Phase III—comparing science ideas to textbooks or other printed resources (for example, writing group notes in response to focus questions). 7. Negotiation Phase IV—individual reflection and writing (for example, creating a presentation such as a poster or report for a larger audience). 8. Exploration of post-instruction understanding through concept mapping, group discussion, or writing a clear explanation.	1. Beginning ideas—What are my questions? 2. Tests—What did I do? 3. Observations—What did I see? 4. Claims—What can I claim? 5. Evidence—How do I know? Why am I making these claims? 6. Reading—How do my ideas compare with other ideas? 7. Reflection—How have my ideas changed? 8. Writing—What is the best explanation that clarifies what I have learned?

Figure 1. The two templates for the SWH: the teacher template and the student template.

Current efforts in science education have highlighted the need for writing to learn strategies to be used in science classrooms (Yore, Bisanz, and Hand, 2003). These strategies recognize the value of having students explain what they know in different ways as a means to construct a richer understanding of science. For example, students can be asked to write in a journal article format or be asked to write a textbook explanation for incoming freshman. These strategies are based on incorporating authentic writing tasks that extend the students' need to engage with the ideas of science, rather than seeing writing as exercises involving notetaking, filling in the blank, or completing the sentence (Prain and Hand, 1996). Writing-to-learn tasks incorporate the need for students to deal with science knowledge and the ways of science, and to begin to understand what it means to build science knowledge and the reasoning strategies required to achieve this goal (Hand, Prain, Lawrence, and Yore, 1999). Constructing science knowledge is not a random activity, but it is a purposeful undertaking based upon posing questions, determining claims, and providing evidence. The SWH is an example of this type of writing activity. Wallace (2004) provides an overview of research studies that support writing as a mode of science learning.

The Science Writing Heuristic (SWH) (see Figure 1) consists of a series of steps to help guide activities or experiments that the students do. By having students explicitly write a claim and then back up their claim with evidence, the process improves each student's reasoning about data. Further, the SWH provides teachers with a template of suggested strategies to enhance learning from laboratory activities. As a whole, using both a guided-inquiry format for the experiments and the Science Writing Heuristic format for the laboratory report provides students with opportunities to be involved in authentic science laboratory activities rather than doing traditional "cookbook" activities. Using inquiry and the SWH, students are encouraged to discuss, debate, and negotiate their understandings of the chemistry. The template for doing a SWH laboratory report (see Figure 1) prompts learners to generate questions, claims, and evidence for claims. It also prompts them to compare their laboratory findings with others, including their peers and information in the textbook, the internet, or other sources. The "Reflections" section in the SWH format prompts students to think about how their own ideas have changed during the experience of the laboratory activity. This approach has been shown to be better than having the students complete a "Discussion" section in their laboratory report.

Using the SWH along with inquiry experiments requires teachers or teaching assistants to be actively involved in helping students understand what they are doing and why they are doing it. By having the students write their own "Beginning Questions," and then having students design their own experimental procedure to answer their question(s), students are more likely to tell you or to write what they are doing and why they are doing it. The SWH provides a teacher template. Instructors know in advance that students need to be able to write a question, design an experiment to answer their question, collect data, analyze the data, and then defend their claims with a series of chain–of-evidence statements that will support the claim. Students defend their claims and evidence to their peers in groups. By using this teaching technique, we are more likely to develop students' understanding of science. When students can justify their understandings, it becomes apparent that they have learned something and are not giving a rote answer. Using guided-inquiry experiments and the SWH laboratory report style achieves this goal in a much better way than traditional experiments using traditional laboratory reports. Students generally spend the least amount of time they can in the laboratory and then complete the writing of their reports at home. Little debate or negotiation occurs within this traditional laboratory setting or even in the write up. The inquiry and SWH activities require that students build scientific argument by having to work with their peers to determine what questions are worth answering in the laboratory, what is the best experimental design to answer the questions, and what claims can be made on the basis of the data collected. This process is part of every SWH laboratory activity. At the conclusion of an experiment, students are required to have whole-class discussions about the outcome of the laboratory activity. The students do this by building their ideas through discussion and negotiation with their peers. The teaching assistant or instructor facilitates the discussion, but does not tell students what they should have learned from the experiment.

In other words, the SWH is designed to promote classroom discussion where students' personal explanations and observations are tested against the perceptions and contributions of the broader group. Students take charge of the discussion and are provided minimal (if any) guidance from the instructor. Learners are encouraged to make explicit and defensible connections between questions, observations, data, claims, and evidence. When students state a claim for an investigation, they are expected to describe a pattern, make a generalization, state a relationship, or construct an explanation. Such activities promote a much greater sense of ownership from the students—they are not simply completing this week's recipe to please the teaching assistant and get the grade they want or need. The knowledge being discussed is their knowledge—they are required to test their knowledge and understanding against the expectations of the activity. Wallace and Hand (2004) provide an overview of the benefits and limitations of using the SWH in the science laboratory.

In a traditional approach, laboratory work often follows a narrow teacher agenda that does not allow for broader questioning or more diverse data interpretation. When procedures are uniform for all students, where data are similar, and where claims match expected outcomes, then the reporting of results and conclusions often lacks opportunities for deeper student learning about the topic or for developing scientific reasoning skills. To address these issues, the SWH is designed to help students think about the relationships between questions, evidence, and claims. The SWH promotes students' participation in laboratory work by requiring them to frame questions, propose methods to address these questions, and carry out appropriate investigations. Such an approach to laboratory work is advocated in many national science curriculum documents on the grounds that this freedom of choice will promote greater student engagement and motivation with topics (for example, NSES, 1996). Does this approach work with college freshmen? The answer is a resounding *Yes*!

Differences Between Traditional and SWH Laboratory Activities

There are a number of differences between traditional and SWH laboratory activities. Guided-inquiry is the main approach used in SWH laboratory activities. Recognizing that students need to understand all the safety requirements and any laboratory techniques where required, safety instructions and laboratory techniques are explicitly taught. However, explicit detailed instructions for doing an experiment are not provided to students. For example, consider a standard calorimetry activity (Figure 2).

The intention of the SWH is to engage students in a process where they are expected to have input and a sense of control over the activities that they have to undertake. **This does not mean the laboratory instructor lets**

the students do what they want—the instructor is required to promote student discussion, both in small and whole class groups, to determine a course of action to achieve the outcomes required.

Traditional	SWH
In a thermos bottle with the lid on, mix 50 mL of 1.0 M HCl with 50 mL of 1.0M NaOH, measure T_i and T_f. Calculate the ΔH_{rxn}. Compare your experimental value of ΔH_{rxn} with the "known" value in the literature.	Design an experiment to see what effect the amount of 1.0 M KOH(aq) solution has on the temperature of the solution when added to 1.0 M HCl(aq). Record time versus temperature data and determine ΔT for each run (minimum of three). Test the solution with pH paper. Compare the time versus temperature graphs for each run. What differences exist among them? Offer a brief explanation. Calculate the number of moles and q. Compare your data with others. Identify a pattern. Identify a relationship among the heats of reaction.

Figure 2. A comparison of a traditional laboratory procedure with an SWH procedure.

Students are required to determine a beginning question that will frame their investigation, that is, they are required to write a question that can be answered by doing the experiment. Students are provided with two exercises, using a whole class discussion and directions provided, that help them write these questions. Initial student attempts at writing researchable questions are quite poor. This is in part because some questions cannot be answered by doing an experiment, while others are trivial. However, they do become better as students adjust to the expectations and demands made of them by the instructor. Figure 3 shows some examples of students' beginning questions from a laboratory activity on calorimetry and thermochemistry.

Non-productive Beginning Questions	Good Beginning Questions
1. What is the atomic weight of my unknown metal? 2. What is the difference between an exothermic and an endothermic reaction? 3. How do you dissolve a salt in solution? 4. Do all acid-base reactions give off heat? 5. What is the difference between ΔH and q? 6. What is the change in enthalpy of my acid-base reaction?	1. Is there a relationship between the atomic weight of metals and the specific heat of metals? 2. What is the most productive ratio of NaOH and HCl that has the greatest ΔH as an exothermic reaction? 3. Is there a relationship between the ΔH, q, and the amount of acid and base that reacts? 4. Do all acid-base reactions have the same ΔH?

Figure 3. A comparison of non-productive and good beginning questions.

Claims and Evidence

We have placed importance on students having to demonstrate scientific argumentation and reasoning skills by ensuring that they are able to state a claim and provide clearly reasoned evidence for their claim. Students will not develop understanding through simply turning to the textbook and copying the answer that they were supposed to obtain for the laboratory activity. They need to actively discuss their investigations, struggle through the analysis process, and be able to derive a claim from what they have investigated. Each group of students is required to put their results on the blackboard so that the entire class can have a chance to examine the different results obtained by the group as a whole. This is important because each group will not be replicating the exact same procedure. Groups can elect to vary the ratio/quantities of materials used, which results in the need for students to examine patterns or trends arising from the experiment. Students will often make a graph to see whether there is a relationship between variables and whether there appear to be any anomalies.

While the students are encouraged to participate in the whole class discussion at the end of the activity to examine the emerging trends or patterns, they are required to individually complete the SWH template as the format of report. Examples of a student's claim and evidence from the calorimetry and thermochemistry activities are given below.

Experiment A. Determination of the specific heat and the atomic weight of a metal.
   Claims: The atomic mass of the unknown metal is 33.9 g/mol. NaOH reacts completely with HCl in a 1:1

ratio. A reaction that gives off energy is exothermic and a reaction that uses energy is endothermic. Dissolving a salt in a solution of water will show a temperature increase or decrease in the solution if it goes through a reaction..

Evidence: We found the heat for the solution using $q = mc\Delta t$. The mass of the water was 50.0 g, the specific heat of water was 4.18 J/g°C, and the change in temperature was 5.80°C. When we multiply the numbers, the heat added to the water was 1212.25 J. We now know that since the water gained heat energy, the reaction gave off energy. This can be written $q_{solution} = -q_{reaction}$. So the heat of reaction was –1212.25 J. We can use this information to calculate what the specific heat of the metal is by using $q/m\Delta t = c$. After looking at the periodic table, we see that the number resembled zinc.

Experiment B. Determination of the change in enthalpy for dissolving a salt in water.
Claim: We proved that we could find out whether or not a salt is endothermic or exothermic based on ΔT. Also, endo/exo cannot be determined by small temperature changes.

Evidence: We tested $MgCl_2$, NH_4Cl, $NaCl$, and $CaCl_2$ at approximately 1 and 2 grams, and regardless of mass, it still showed relatively the same change in temperature. Calcium chloride was the best one because when we doubled the mass, the temperature also doubled.

Experiment C. Determination of change in enthalpy of dissolving in water.
Claim: By adding more mass of $MgSO_4$, the q_{rxn} and ΔT increase, but ΔH_{rxn} stays the same in proportion to each mass.

Evidence: For each experimental run, the mass and ΔT for the solution were measured. We used $q = mc\Delta T$ to calculate q. The q values were different, but when you divide q by the number of moles, you get ΔH. By keeping the total mass constant (mass of salt + mass of water = total mass), when you add more mass of salt, you get a bigger ΔT and bigger q. When you increase the mass of salt, you increase the number of moles. We get ΔH by dividing q by the number of moles.

Experiment D. Determination of the heat associated with an acid-base reaction.
Claim: When doing the acid-base reactions, you must take into account the limiting reagent. The ratio of reaction of acid to base is 1:1, according to the equation, but if you have more moles of one than the other, then that is present in excess. It is not simple. The q_{rxn} changes but the ΔH does not.

Evidence: As shown in our graph of ΔT versus volume of HCl (Figure 4), HCl is the limiting reagent. Keeping the number of moles of NaOH constant, as we add HCl, the ΔT increases up to a point, then it decreases. There is a relationship between the number of moles of acid and base that react and the change in temperature and q. The graph shows that after all of the base reacts, when we add more acid, the temperature does not change because there is no longer a chemical reaction that occurs. The excess mass of the acid is just more mass that absorbs the heat generated, so the total solution cannot achieve a greater ΔT.

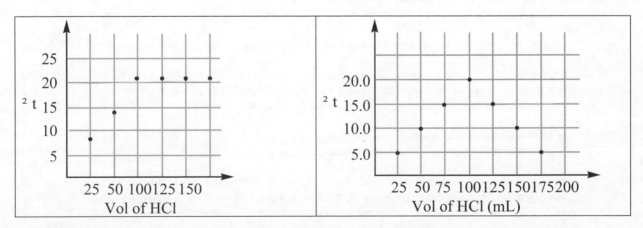

Figure 4. Student graphs of ΔT versus volume of HCl (aq) added for a calorimetry experiment.

The Science Writing Heuristic and Encouraging Students to Design Experiments

In organizing the SWH laboratory work to be completed (as described above), the students are required to be much more active in designing the activity than with the more traditional format. In a guided-inquiry Science Writing Heuristics format, students are not explicitly told how to do each experiment. We have found that students who are engaged in doing the SWH approach, with practice, are more apt and able to design an experiment to answer a specific question than students who do a traditional laboratory activity. They are more able to evaluate the process they performed, as well as propose changes to the existing procedure (Berg, Bergedahl, Lundberg, and Tibell, 2003). An example of this control over the activity is illustrated by the following quote from a student:

"We designed an experiment to calculate the best ratio of HCl and NaOH so that all of the moles of each react completely. To do this, we changed both the HCl and NaOH volume and measured the change in temperature. We hoped to see a graph that curved upwards, but it did not."

"In our second experiment, we changed the volume of HCl but kept the volume of NaOH constant. By doing this, we saw what ratio of HCl works best to give off the most heat. We calculated q, the heat of reaction, and we calculated ΔH of the reaction. We contributed our data to the class data. One thing unexpected was, despite using different acids, everyone's acid-base ΔH was about the same."

Students' Understanding of Chemistry Concepts Improves as a Result of Using the Science Writing Heuristic

Student performance on laboratory practical examinations. We now require students to write explanations about chemistry events as part of the laboratory practical examination. We believe it is necessary for the reasoning and argumentation skills required of the students during their weekly laboratory activities to be carried across to the practical examination. A typical practical examination task is described In Figure 5.

You will be given a salt. Set up and run an experiment that will allow you to calculate the ΔH for the salt dissolving in water. We will call this ΔH_{rxn}.

a. $\Delta H_{rxn} =$.
b. Write a balanced equation for the dissolving process. Include states of matter with each species.
c. Record all data and observations.
d. Classify this reaction as endothermic, endothermic, or neither. Justify your response.
e. What is gaining heat? What is losing heat?
f. Write a brief explanation of how the dissolving process of your salt generates heat, absorbs heat, or does not exchange heat.

Figure 5. A sample laboratory practical task on thermochemistry.
Sample student answers for part (f) from the practical exam task on thermochemistry are discussed below:

I.   Correct or nearly correct written explanations from two students having an effective SWH instructor.

"NaCl dissolving in water is an endothermic process because the solution became cooler when the salt dissolved. When the salt is added to water, the NaCl and H_2O mix and absorb energy because it takes energy to break bonds and when bonds form energy is released."

"When KCl dissolves in water, the reaction is endothermic. The temperature of the water dropped because the reaction took in some of its heat."

Notice that each student has written an explanation that correctly identifies a decrease in the temperature of the solution or water with the dissolving process. They identify the dissolving process as an endothermic process, because energy was required by the dissolving process.

II.   Incorrect or mostly incorrect written explanations from three students having an ineffective SWH instructor.

"When KCl dissolves in water, it is an exothermic process because the solution got colder and produced a negative ΔH. The solution gets colder because it absorbs heat from the outside air."

"NaCl dissolving in water is an exothermic process because the temperature inside of the calorimeter dropped and made the reaction exothermic. The salt absorbed some of the heat from which the water had."

"$MgSO_4$ dissolves in water to make an exothermic process. It generates heat because the process causes the solution to increase in temperature."

Notice that in each of these explanations, the students have incorrectly identified the dissolving of NaCl or KCl as an exothermic process. The first two students still have the mental model of a physical process they are using. They appear to focus on object A decreasing temperature and object B increasing temperature. The first student incorrectly associates all processes that absorb heat with an increase in temperature. This student cannot visualize energy being transferred into a system (in this case, a dissolving process) and not have it increase in temperature. Greenbowe and Meltzer (2003) reported on the difficulties students have with calorimetry and thermochemistry problems. Burke, Pook, Cantonwine, Greenbowe, and Hand (2003) reported that students and instructors using the SWH approach were able to overcome most of these difficulties with thermochemistry.

Student Performance on Lecture Examinations and on ACS Examinations. In order to investigate the effect that SWH has on students' performance on lecture examinations, Greenbowe, Poock, Burke, and Hand (2003) designed an experiment comparing students in an effective SWH laboratory section (both the teaching assistant and the students were effectively implementing SWH) to students in an ineffective laboratory section (both the teaching assistant and the students were not effectively implementing SWH). Two independent observers rated both the students and the instructors in a general chemistry laboratory over a three-month period. Figure 6 shows the major differences between ineffective and effective instructors.

On the American Chemical Society California Diagnostic test, at the beginning of the semester, there was no significant difference ($F(1, 285) = 1.1269, p = .289$) between students in sections taught by different instructors. Upon completion of the laboratory course, there was a significant difference on examinations given in the lecture component of the course ($F(1, 236) = 8.7204, p = .004$) with higher performance in sections taught by high-implementing SWH instructors than in sections taught by low-implementing SWH instructors. An effect size (Cohen's d) of 0.38 was obtained.   Figure 7 shows the average scores of students on the ACS California Diagnostic Test, the comprehensive course final exam, and on the ACS First Semester General Chemistry Examination as a function of the type of instructor and the type of students they had in the laboratory.

Ineffective Instructor	Effective Instructor
Beginning questions not discussed.	Opportunities to discuss beginning questions.
TA tells students exactly what needs to be done. Individual work or pairs work separately from the class. Instructor assigns tasks.	Setting up the lab for student-centered work.
No sharing or analysis of class data.	Allowing students to assign groups and tasks.
Students immediately leave when finished with their work.	Class data are presented on the chalkboard. Class data are analyzed as a group. Instructor guides a class discussion of concepts covered in the laboratory.

Figure 6. A comparison of ineffective instructors to effective instructors.

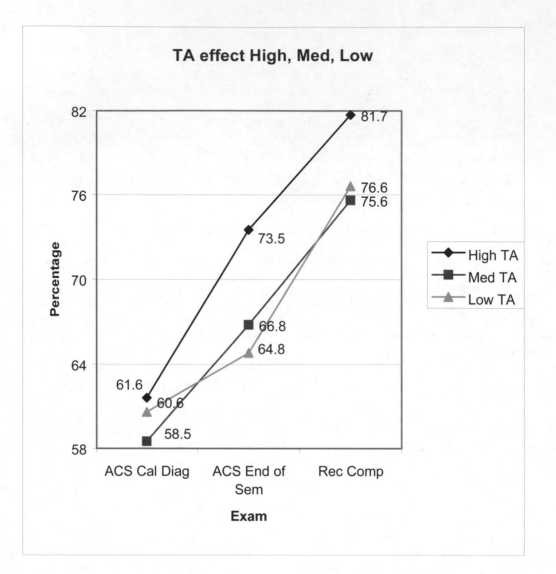

Figure 7. A comparison of average scores of students based upon the classification of the type of teaching assistants guiding their sections.

The results of this research present evidence that there is a connection between effective teaching and learning in the chemistry laboratory (inquiry/SWH) and student performance in lecture and on a standardized ACS examination.

The implementation of the SWH approach was beneficial particularly for females, and was able to close the gender gap. At the beginning of the course, there was a significant difference in favor of males ($F(1, 285) = 14.5298$, $p < .001$) on the ACS California Chemistry Diagnostic Test. At the completion of the semester, there was no significant difference between males and females on the ACS General Chemistry First Semester Examination ($F(1, 236) = .0822$, $p = .775$). The effect size due to gender was reduced from 0.45 to 0.04 from the start to the end of the semester. Figure 8 shows the average scores of male and female students on the ACS California Diagnostic Test, the final exam, and on the ACS First Semester General Chemistry Examination.

Figure 8 shows the average scores of male and female students on the ACS California Diagnostic Test, the final exam, and on the ACS First Semester General Chemistry Examination. Figure 9 shows the performance of male and female students as a function of the effectiveness of their instructor with respect to the implementing the Science Writing Heuristic in the laboratory.

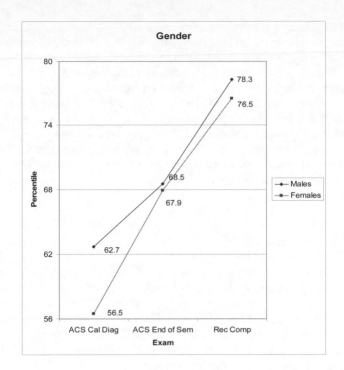

Figure 8. Comparison of scores on chemistry exams for male and female chemistry students.

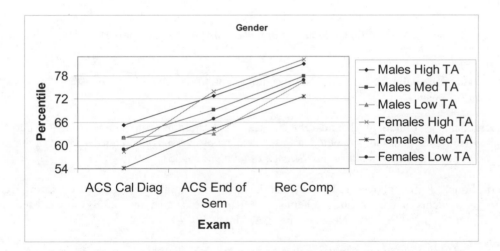

Figure 9. Comparison of scores on chemistry exams for male and female chemistry students as a function of the effectiveness of their laboratory instructor.

In this study at the beginning of the course, females started at a lower level of chemistry knowledge compared with the males in the course with respect to the average scores on the ACS California Chemistry Diagnostic Test. Females were able to "catch-up to" their male counterparts, in part, by having instructors in the chemistry laboratory who effectively implemented the Science Writing Heuristic with the guided-inquiry experiments.

Summary

This chapter presents an introduction to some of the models of writing to learn science and the Science Writing Heuristic. Writing is an important component of the learning process and an important component of assessing students' understanding of chemistry. Students can gain insights and develop a deeper understanding of science content through structured writing assignments writing, such as the Science Writing Heuristic. Chemistry instructors tend to rely on answers to numerical problems (or multiple-choice problems) to assess understanding of chemistry. Students can generate correct numerical answers to problems, but often they do not have a real understanding of what these numbers represent (Nurrenbern and Pickering, 1987; Sawrey, 1990). By using guided-inquiry, small-group work, and the Science Writing Heuristic as the basis of the laboratory activity and the laboratory report, students do better on lecture examinations and laboratory practical tasks. Females benefit by having an effective SWH instructor in that their scores on examinations improve more so than males. Other research studies (Keys, Hand, Prain, and Collins, 1999; Hand, Hohenshell, and Prain, 2004) provide evidence that the SWH format for structuring laboratory activities and notebooks helps students understand science concepts better than students who used a traditional laboratory notebook format. The SWH is an important component of doing inquiry in the chemistry laboratory and facilitates student thinking about chemistry. While there is a period of adjustment for both students and instructors to implement inquiry and the SWH, we believe that the benefits to be gained are very worthwhile.

Suggested Reading

Rudd, J. A., Greenbowe, T. J., and Hand, B. M. (2002). Re-crafting the general chemistry laboratory report. *Journal of College Science Teaching.* 31(4), 230–234.
> Provides a brief overview of the science writing heuristic as it relates to its incorporation into a college-level general chemistry laboratory course.

Burke, K. A.; Hand, B. M.; Poock, J.; and Greenbowe, T. J., (*in press*). Training Chemistry Teaching Assistants to Use the Science Writing Heuristic. *Journal of College Science Teaching.*
> Practical suggestions for training teaching assistants, instructors, and students in the Science Writing Heuristic in chemistry laboratory sections.

References

Abraham, M., and Pavelich, M. (2000). *Inquiries into chemistry, 3ʳᵈ ed.* Prospect Heights, IL: Waveland Press..

Bereiter, C., and Scardamalia, M. (1987). *The psychology of written composition.* Hillsdale, NJ: Lawrence Erlbaum Associates.

Berg, C., Bergedahl, V., Lundberg, B., and Tebell, L. (2003). Benefiting from an open-ended experiment? A comparison of attitudes to, and outcomes of, an expository versus open-ended version of the same experiment. *International Journal of Science Education.* 25, 351.

Burke, K. A., Poock, J., Greenbowe, T. J., and Hand, B. M. Using inquiry and the Science Writing Heuristic to improve students' understanding of chemistry. A paper presented at the 227th National Meeting of the American Chemical Society. New Orleans, LA. March, 2004.

Connally, P. (1989). *Writing to learn mathematics and science.* NY: Teachers College Press.

Elbow, P. (1973). *Writing without teachers.* London: Oxford University Press.

Emig, J. (1977). Writing as a mode of learning. *College Composition and Communication.* 28, 122–128.

Flower, L. S., and Hayes, J. R. (1980). Images, plans, and prose: The representation of meaning in writing. *Written Communication.* 1, 1, 102–160.

Galbraith, D. (1999). Writing as a knowledge-constituting process. In M. Torrance and D. Galbraith (Eds.), *Knowing what to write: conceptual processes in text production.* 139–160. Amsterdam: Amsterdam University Press.

Galbraith, D. (1996). Self-monitoring, discovery through writing and individual differences in drafting strategy. In G. Rijlaarsdam, H. van den Bergh, and M. Couzijn (Eds.). *Theories, models, and methodology in writing research.* 121–141. Amsterdam: Amsterdam University Press.

Greenbowe, T. J. and Meltzer, D. A. (2003). Student learning of thermochemical concepts in the context of solution calorimetry. *International Journal of Science Education*. 25(7), 779–800.

Greenbowe, T. J., Poock, J., Burke, K. A., and Hand, B. M. (October 27, 2003). Training teaching assistants to teach using inquiry and the science writing heuristic yields high dividends for students. A paper presented at the 59th Southwest Regional Meeting of the American Chemical Society. Oklahoma City, OK.

Gunstone, R. F., and Champagne, A. B. (1990). Promoting conceptual change in the laboratory. In . E. Hegarty–Hazel (Ed.) *The student laboratory and the science curriculum* 159–182. London: Rutledge.

Hand, B. (2004). Cognitive, constructivist mechanisms for learning science through writing. In C. Wallace, B. Hand, and V. Prain. *Writing and learning in the science classroom.* Dordrecth, the Netherlands: Kluwer Academic Publishers.

Hand, B. M., Hohenshell, L., Prain, V. (2004). Exploring students' responses to conceptual questions when engaged with planned writing experiences: A study with year 10 science students. *Journal of Research in Science Teaching*. 41(2), 186–210.Hand, B., Prain, V., Lawrence, C., and Yore, L. D. (1999). A writing in science framework designed to improve science literacy. *International Journal of Science Education*. 10, 1021–1036.

Herron, J. D., (1996). *The Chemistry Classroom.* Washington, DC: ACS.

Herron, J. D., Greenbowe, T. J. (1986). What can we do about Sue: A case study of competence. *J. Chem. Ed.* 63(6), 528–531.

Hofstein, A., and Lunetta, V. N. (1982). The role of laboratory in science teaching: Neglected aspects of research. *Review of Educational Research*. 52, 201–217.

Inquiry and the NSES: A Guide for Teaching and Learning. (2000). Washington, DC: National Academy Press..

Keys, C. (1999). Revitalizing instruction in scientific genres: Connecting knowledge production with writing to learn in science. *Science Education*. 83(2), 115–130.

Keys, C., Hand, B., Prain, V., and Collins, S. (1999). Using the science writing heuristic as a tool for learning from laboratory investigations in secondary science. *Journal of Research in Science Teaching*. 36, 10, 1065–1084.

Langer, J., and Applebee, A. (1987*). How writing shapes thinking: A study of teaching and learning.* (NCTE Research Rep. No. 22). Urbana, IL: National Council of Teachers of English.

Lawson, A., and Renner (1989). *A Theory of Instruction: Using the Learning Cycle.* NARST Monograph #1.

Lazarowitz, R., and Tamir, P. (1990). Research on using laboratory instruction. In D.L. Gable (Ed.) science. *Handbook of Research on Science Teaching and Learning*. NY: MacMillian

Landis, C., Ellis, A., et al. (2001). *Chemistry ConcepTests.* Upper Saddle River, NJ: Prentice Hall.

Nakhleh, M., Polles, J., and Malina, E. (2002). Learning chemistry in the laboratory environment. In J. Gilbert, O. De Jong, R. Justi, D. Treagust, and J. Van Driel (Eds.) *Chemical education: towards research-based practice*. Boston: Kluwer Academic.

Nakhleh, M., and Krajcik, J. (1990). A protocol analysis of the influence of technology on students' actions, verbal commentary, and thought processes during the performance of acid-base titrations. *Journal of Research in Science Teaching*. 30(9), 1149–1168.

Nakhleh, M. (1992). Why some students don't learn chemistry. *Journal of Chemical Education*, 69, 191–196.

Nakhleh, M., and Krajcik, J. (1994). Influence of levels of information as presented by different technologies on students' understanding of acid, base, and pH concepts. *Journal of Research in Science Teaching*. 31(10), 1077–1096.

National Science Education Standards. (1996). Washington, DC: National Academy of Sciences Press.

Nurrenbern, S. C., and Pickering, M. (1987). Concept learning versus problem solving: Is there a difference? *Journal of Chemical Education*. 64(6), 508–510.

Pickering, M. (1987). Laboratory education as a problem in organization. *Journal of College Sciene Teaching*. 16, 187-189.

Poock, J., Burke, K.A., Cantonwine, D., Greenbowe, T. J., and Hand, B. M. (March, 2003). "Evaluating the effectiveness of implementing inquiry and the Science Writing Heuristic in the general chemistry laboratory: Teaching assistants and students." New Orleans, LA.: ACS.

Prain, V., and Hand, B. (1996). Writing for learning in secondary science: Rethinking practices. *Teaching and Teacher Education*. *12*(6), 609–626.

Rudd, J. A., Greenbowe, T. J., and Hand, B. M. (2002). Re-crafting the general chemistry laboratory report. *Journal of College Science Teaching*. 31(4), 230–234.

Rudd, J., Greenbowe, T. J., and Hand, B. (2001). Using the science writing heuristic to move toward an inquiry-based laboratory Curriculum. *Journal of Chemical Education*. 78(12), 1680–1686.

Sawrey, B. A. (1990). Concept learning versus problem solving: Revisited. *Journal of Chemical Education*. 67(3), 253–254.

Stake, R.E., and Easley, J. (1978). Case studies in science education. Urbana–Champagne, University of Illinios, Center for Instructional and Curricular Evaluation.

Tynjala, L., Mason and K. Lonka (Eds.) (2001). *Writing as a learning tool: Integrating theory and practice*. (105–129) Dordrecht, the Netherlands: Kluwer Academic Publishers.

Tobin, K., and Gallagher, R. (1987). What happens in high school classrooms? *Journal of Curriculum Studies*. 19, 549–560.

Wallace, C. S. (2004). Evidence from the literature for writing as a mode of science learning. In C. S. Wallace, B. Hand, and V. Prain (Eds.) *Writing and learning in the science classroom*. Boston: Kluwer Academic Publishers.

Wallace, C. S. and Hand, B. (2004). Using a science writing heuristic to promote learning from the laboratory. In C. S. Wallace, B. Hand, and V. Prain (Eds.) *Writing and learning in the science classroom*. Boston: Kluwer Academic Publishers.

Yore, L. D., Bisanz, G. L., and Hand, B.M. (2003). Examining the literacy component of science literacy: 25 years of language and science research. *International Journal of Science Education*. 25, 689–725.

13

Team Learning

Pratibha Varma-Nelson
Department of Chemistry
Northeastern Illinois University

Brian P. Coppola
Department of Chemistry
University of Michigan

Abstract

Team Learning is a second-generation pedagogy; that is, it results from a combination of robust and well-established theories and methods for instructional design. Team Learning is presented here as a consequence of a four-part framework—one that combines group (collaborative and cooperative) learning, reciprocal teaching, Vygotsky's educational theory, and studio instruction. Two models of team learning in chemistry are presented: Peer-Led Team Learning (PLTL), a successful model that has been replicated in many venues and in other disciplines; and Structured Study Groups (SSG), which provide a student-led Honors option for students in a large, introductory chemistry program.

Biographies

Dr. Pratibha Varma-Nelson is Professor of Chemistry and Chair of the Department of Chemistry, Earth Science, and Physics at Northeastern Illinois University, Chicago. She received her B.Sc. in Chemistry with first class in 1970 from the University of Pune, India, and a Ph.D. in 1978 from the University of Illinois in Chicago in Organic Chemistry. The title of her thesis was "Protein Ancestors: Heteropolypeptides from Hydrogen Cyanide and Water." She completed a Post Doctoral fellowship in Enzymology at the Stritch School of Medicine, Loyola University, Maywood, Illinois, before joining the faculty of Saint Xavier University, Chicago, in 1979. At SXU, she taught courses in Organic Chemistry and Biochemistry. At NEIU, she teaches a Capstone Seminar to students majoring in chemistry, and she also teaches Chemistry of Biological Compounds.

Since 1995, her professional activities have revolved around the development, implementation, and dissemination of the Peer-Led Team Learning (PLTL) model of teaching. She was an active partner in the Workshop Chemistry Project, one of the five NSF-supported systemic reform projects in chemistry, and she is currently a co-PI of two NSF-supported National Dissemination Grants: "Peer-Led Team Learning: Building the National Network" and "Multi Initiative Dissemination." In addition, she has co-authored several publications and manuals about the PLTL model. Dr. Varma-Nelson is director of the Workshop Project Associate (WPA) Program, which provides small grants to facilitate the implementation of PLTL, and director of the annual Chautauqua course on PLTL.

Dr. Brian P. Coppola is *Arthur F. Thurnau* Professor of Chemistry at the University of Michigan, Associate Chair for Curriculum and Faculty Affairs, and a Faculty Associate at the University of Michigan Center for Research on Learning and Teaching. He received his B.S. degree from the University of New Hampshire in 1978, joined the faculty at the University of Wisconsin–Whitewater in 1982, and received his Ph.D. in Organic Chemistry from the University of Wisconsin–Madison in 1984.. When he moved to Ann Arbor in 1986, Dr. Coppola joined an active group of faculty in the design and implementation of a revised undergraduate chemistry curriculum. His recent publications range from mechanistic organic chemistry research in 1,3-dipolar cycloaddition reactions to educational philosophy, practice, and assessment.

In 1998, Dr. Coppola was selected as part of the first group of Carnegie Scholars affiliated with The Carnegie Foundation for the Advancement of Teaching's CASTL program (Carnegie Academy on the Scholarship of Teaching and Learning). He also currently directs the CSIE program (Chemical Sciences at the Interface of Education), which broadens the scholarly development for all students (undergraduate to post-doctoral) who are interested in academic careers. These students have the opportunity to collaborate with the faculty on teaching projects in the same way that they pursue research projects. The centerpiece of the program is built on the familiar structure of a graduate training grant that provides a novel opportunity for the department's Ph.D. students who wish to add this work to their theses.

Introduction

Knowledge is constructed in diverse ways, following strategies and traditions evolved by the disciplines over time. One way knowledge can be constructed is by individuals working in relative isolation and intense contemplation. This method is probably more familiar to our colleagues outside of the science, technology, engineering, and mathematics (STEM) disciplines than it is to us. Faculty in the STEM disciplines have devised an alternative method of knowledge construction as a response to the increasingly sophisticated demands of modern research. They have refined a system that distributes the responsibility for getting work done over an intergenerational team, namely, the research group (Coppola, 1996; NSF). The faculty advisor is also the research director, and orchestrates the activities, of the research group. The roles of research director and advisor are critically intertwined, because this person oversees (manages) the overall direction of the team, while simultaneously balancing the challenges and skill requirements of tasks and matching them with team members in order to fulfill the dual goals of productivity and education. Hoffmann (2003) believes the learning opportunities provided by research groups, as we know them, are the primary reasons why graduate students in the United States leapfrog in creative ability over their better-trained European and Asian counterparts during graduate school. The scientific research group is an excellent model for an effective team. The structure of these teams is quite sophisticated because the team (a collection of roles and a process) remains in place, while novices and other new members flow in as trained experts depart. When the Ph.D. is awarded, an individual has stepped through the multiple roles within the team during the educational process, until (based on primarily self-generated results) that person is an expert (an authentic leader) on a narrow body of work.

Regardless of the level, teams work: they educate well and they educate efficiently by structuring and supporting the time a person spends on a task. In this chapter, we argue that progress in understanding designs for teaching and learning is following a parallel path to research and is responding to the same pressure—that is, increased sophistication and the need to do better with more demands within the constraints of finite time and limited resources. Over a relatively short time, instructional design for classroom learning—another version of orchestrating the construction of knowledge—has systematically shifted towards recognizing the value of formally distributing responsibility for teaching and learning by integrating group strategies as a part of effective instructional practices.

Team Learning (or TBL: "Team-Based Learning") has a rich history derived from extensive work in the pre-college arena. Larry K. Michaelsen is generally credited with the generative production and application of these ideas in higher educational settings (Michaelsen, 2004). Team learning positively impacts student learning because it draws from and combines four well-established areas of educational design and research. We believe that this provides a useful theoretical framework for thinking about Team Learning that has been historically absent from the practical recommendations provided by Michaelsen and others (Coppola, 2004). These recommendations are: (1) group (collaborative and cooperative) learning, (2) reciprocal teaching, (3) Vygotsky's educational theory, and (4) studio instruction. Finally, we also agree with those critics who remind us that pedagogies should be intimately aligned with instructional goals rather than treated as universal truths (or "magic bullets") that represent exclusive solutions (Coll and Taylor, 2001). In other words, group learning strategies should complement and strengthen other instructional strategies, including the lecture, and all strategies should be exploited for the educational gains that can be best realized through them.

Without a doubt, one of the drivers in undergraduate science education reform has been the inverse relationship between domestic students choosing scientific and technical career pathways as the projected needs of a more scientifically literate and technically proficient workforce are increasing. Not only are graduation rates for students in STEM fields far below national averages for other disciplines, but this is an especially acute problem for underrepresented minorities. One proposition is that, without effective intervention and support, many students decide not to continue in STEM majors or careers. A study by Astin and Astin (1993) reported that 40% of first-year college students majoring in STEM fields switched majors before their senior year. Another study by Astin (1993) outlined several factors that positively affect the quality of intellectual development during college years. The ones that are most relevant to the present discussion are: (1) student faculty interaction outside the classroom, (2) involvement with student peer groups, (3) involvement on campus through various forms of community-building activities, and (4) the amount of time that students devote to studying, i.e.,"time on task," including the quality of time spent based on the design of the tasks and how well mechanisms of support (scaffolds) mediate difficulties as they arise. Of these factors, Astin (1993) identified "the student's peer group as the most potent source of influence on growth and development during the undergraduate years."

These studies point to a need for using pedagogies that tap into the benefits of collaborative learning techniques. As argued above, scientists should be able to adapt easily to this way of thinking because science relies so strongly on these same methods—although it requires a significant shift in thinking, namely, that the teaching that one orchestrates as a research director is metaphorically related to the teaching one orchestrates in classroom settings (Coppola and Pearson, 1998). In addition, as noted by Howard Hills, we have reached a stage where, although "technology will provide more and more learning support," interacting with technologies also reduces the benefits from face-to-face social interaction in building knowledge. Thus, "this collaborative team-learning approach retains the human contact we all need" (Hills, 2001). In this chapter, we review the attributes of team learning as a form of group learning, summarize the characteristics of an effective team leader, and suggest some specific pedagogical benefits from organizing team activities as a part of instructional design. We will then provide some brief design features based on our experiences in implementing team learning activities in two specific programs. In both cases, we are defining teams as groups of students in undergraduate courses who are led by another undergraduate student who not only has prior experience in the course, but who has also been identified for and trained in the requisite leadership skills to transform a group into a high performing team.

The Nature of Teams and Team Learning

What is the difference between a group and a team? Our colleagues in Business and Business Administration have spent a great deal of time defining these attributes (e.g., Torres and Spiegel, 1991), while those in education have promoted teams as a way to inform the design of learning environments for children (e.g., Atkinson, 2001; Michaelsen et al., 2002). In translating these ideas for higher education, Michaelsen writes about the transformation of a group to a team in order to produce effective learning: "The key to successful use of team-based learning is knowing how to transform groups into teams which then are capable of producing powerful learning."

A group is a gathering of people who can end up with a variety of organizational characteristics; one of these organizational sets is a team. Katzenbach and Smith define "team" as a small number of people with complementary skills who are committed to a common purpose, a set of performance goals, and an approach for which they hold themselves mutually accountable (1993). They also stated the importance of ground rules, which then become a part of the culture of the team. Their taxonomy of characteristic differences between teams and groups is a useful framework for thinking about the changes that take place during the transformation of a group to a team.

Hills (2001) observes four stages in the development or formation of teams from groups. These include excitement, grousing, confusion, and performance. The final stage (performance) is the desirable outcome; shortening the time it takes to work through the first three stages is key to the success of Team Learning. The single most important determinant for making progress as a team is the presence of a leader with sufficient preparation and training. A leader may be an undergraduate student who has previously taken the course, a graduate student, or a faculty member. Good leaders can quickly get their teams to the performance stage, where they are effective and efficient without being dependent on the leader.

In describing the five dysfunctions of a team, Lencioni (2002) simultaneously identifies the important organizational goals for building a team from a group: (1) developing mutual trust, (2) creating mechanisms to resolve conflicts, (3) becoming committed to each other in accomplishing tasks, (4) accepting accountability, and (5) providing attention to quality results. The following characterization of teams and groups represents extremes on a continuum.

Characteristics of Teams	*Characteristics of Groups*
share leadership roles as they see fit	have a strong and clearly focused leader
take mutual, collective accountability	take individual accountability
create specific team objectives that they deliver themselves	have imposed, mandated, granted and/or the same objectives as the broader organizational mission
deliver collective products/outcomes	deliver individual products/outcomes
encourage open-ended discussion and active problem–solving meetings	run efficient meetings
measure performance directly by assessing collective products/outcomes	measure effectiveness indirectly by their influence on others
discuss, decide, and do real work together	discuss, decide, and delegate
have open and honest dialogue	have polite discussions
have fun working together and laugh a lot	just work
can't wait to be together	meet because it is a requirement

Learning Outcomes

Individuals depend on (and learn with) one another in multiple situations and settings, including, but not limited to, those we design for them. In order to emphasize this idea, Bruffee (1995) advocates a phrase attributed to John Dewey as a hallmark educational goal: learning to live "an associated life." As Bruffee describes it, the unique strength of formal education in the United States has been a philosophical basis of "associated learning" since at least the time of Benjamin Franklin. Team learning prepares students to work effectively with people, including those who are different from themselves; it generally builds reliance on, and trust and respect for, the perspectives and work of others. Including a strong focus on process as well as product ensures that even good students benefit from learning and appreciating diverse approaches. In particular, explanatory and sense-making are skills derived from discussion, critique, and other aspects of reflective practice, such as effective revision and editing of others' work and ideas (Schön, 1983). The simplest proposition is that students have the opportunity to express what they know by participating in a team, and in so doing, learning gains should be reflected in traditional measures, such as test scores.

Some models of team learning rely on students who have successfully completed the course to serve as peer leaders; in other models, faculty themselves may choose to serve as the facilitators of teams. The advantage of using peer leaders is that they carry "the authority of recent experience" with both the university and the course (Coppola, Daniels, and Pontrello, 2001).

A significant outcome from peer-led instruction is the effect on the peer leader. The effective habits of good teachers should emerge and be encouraged by students who take on these roles. In addition to numerous anecdotal reports of increased comfort with the subject matter and confidence with addressing ideas in public settings, recent research findings (Gafney and Varma-Nelson, 2001; Tenney and Houck, 2004) point to the following gains when students take on the role as peer leaders:

- Increased content knowledge and better success in higher-level science courses.
- Increased confidence to pursue science-related careers.
- An appreciation for different learning styles.
- Improved "people skills" and a collegial relationship with the course instructor.

If peer leaders lead the teams, they must be provided with proper training and support so that they are effective facilitators. If team learning is used in introductory classes, especially in a large institutional setting or at an urban commuter institution where one's academic life might seem more easily "dissociated" rather than "associated," student leaders can serve as role models as well as facilitators for the transition of the new students to college life; this may result inan outcome that lasts long after the details of the course content have been diluted by other courses as the years go by.

Team learning requires students to negotiate meaning and understanding in debate and discussion with peers. This is the essential mechanism by which scientists construct knowledge and understanding; it promotes the kind of learning that is good for their future. In industry or in an academic environment, scientists rarely work alone, but they do have individual responsibilities within the context of a larger task. In graduate school, they participate in the various functions of a research group, many of which rely on communicating about and building on each other's work.

How and Why do Teams Impact Learning? A Theoretical Framework

There is nothing in the Team-Based Learning literature that points to a theoretical framework for how and why teams have a positive impact on student learning. Here, we propose that four previously defined contexts interact and inform Team Learning.

The first of these areas is *group learning* (cooperative and collaborative learning methods; see Chapter 11). Collaborative learning follows a social constructivist model of learning, namely, that knowledge is created through interaction between people that builds on prior knowledge, resulting in "critical thinking, problem solving, sense making and personal transformation, the social construction of knowledge-exploration, discussion, debate and criticism of ideas" (Boud, Cohen, and Sampson, 2001). Team learning not only incorporates elements of cooperative learning (Johnson and Johnson, 1999) such as positive interdependence and face-to-face interaction, but also includes a trained team leader who holds each member accountable, facilitates decision making, and provides feedback on how the team is performing.

The second contribution comes from the area of *reciprocal teaching* (Palincsar and Brown, 1984; Brown and Palincsar, 1989) and the power of *explanatory knowledge* (Coleman, 1998; Coleman, Brown, and Rivkin, 1997). In reciprocal teaching, instructional tasks are designed by studying and breaking down the strategies used by successful learners and then are utilized to guide learning by novices. Research on explanatory knowledge concludes that students need to reflect on their learning and develop interpersonal communication skills as a part of understanding. Coleman's results are compellingly clear: a student who anticipates the need to make a subsequent explanation about something which is being learned will learn it better in the first place. In team learning, promoting reflective practice helps students understand that true learning comes from discussing the basis for one's answers and conclusions rather than memorizing the answers (Chambers and Abrami, 1991).

The third area that informs team learning is based on the educational theories attributed to Lev Vygotsky. Vygotsky believed strongly that community was necessary in order for students to "make meaning" (Vygotsky, 1978, 1985). There are two critical components in Vygotsky's model: (1) an individual who has a higher level of ability than the learner with respect to the particular idea or activity that needs to be learned (the MKO, or More Knowledgeable Other), and (2) tasks designed to press the learner to a reasonable expectation of achievement or understanding by providing appropriate assistance (scaffolding) that is easily removed as learning gains are achieved (these are tasks within the Zone of Proximal Development, or ZPD). ZPD is the most important concept for designing appropriate instructional materials. Tasks that are merely drill problems that do not require ideas to be synthesized or creative work to be done fall below the threshold for ZPD tasks; however, applying the ideas derived from reading Lavoisier in the original French to solving structural chemistry problems are likely to be beyond a reasonable ZPD.

The fourth area, *studio instruction,* is the least well examined in science teaching and learning even though it is a part of the implicit framework that describes team learning (Rieber, 2000). The essence of *studio instruction* is to have students generate materials that represent their learning. In the traditional artist's studio, artifacts that represent student learning are writing, painting, musical composition, sculpture, etc. In the research laboratory, experimental results (such as new compounds, new separation methods, etc.) represent products of learning; they are documented and ultimately legitimized by subjecting them to peer review and critique. In a problem-solving team, a student's particular solution to a problem serves as a product of learning. These products that clearly represent student learning then become an object of study, peer review, and critique, and the group learns from the collective understanding derived from individual interpretations of common tasks. In a studio format, tasks need to be sufficiently sophisticated so that errors and misunderstandings emerge as a topic of conversation but are not so demanding that transfer of learning is impossible (Bransford and Schwartz, 1999). Engineering programs have developed sophisticated instructional spaces to support studio instruction (Wilson, 1994), and a few chemistry departments have begun to follow suit (Bailey, Kulinowski, and Paradis, 1998; Apple and Cutler, 1999; Sweeder, et al., 2003; Blunt, et al., 2003; Gottfried, et al., 2003).

Breaking the tradition of [only] centralized authority in teaching and learning coincides with society's demands for increasing the diversity of people who are prepared to do (or understand) science and technology. This is fortunate, because many believe that this increase can be accomplished by designing classrooms that foster success both broadly and inclusively. As a case in point, Seymour and Hewitt (1997) showed that "the most effective way to improve retention among women and students of color, and to build their numbers over the longer term, is to improve the quality of the learning experience for all students—including non-science majors who wish to study science and mathematics as part of their overall education." They also found that while almost all students value collaborative learning, students from underrepresented groups "appreciate it more and miss it when [it is] unavailable." Their research identified interactive collaborative learning as key to improving student performance. Triesman (1992) showed that students often fail to excel in science and mathematics because they do not know how to work effectively with peers to "create a community for themselves based on shared interest and common professional goals." He found that remedial programs and those specifically targeted at minority groups often do not increase success rates. Indeed, Steele (Steele, 2000) has argued convincingly that institutional identification based on deficiency, while well-intended, becomes a self-fulfilling prophecy whose victims "underperform" to prescribed institutional stereotype and bias. Thus, programs designed to improve learning must include all students.

Models for Team Learning in Chemistry

Team learning is not a single model. It covers a range of different but related approaches. Two of these are described in this section.

A. Peer-Led Team Learning (PLTL)

Introduction. The PLTL workshop model (Gosser and Roth, 1998; Sarquis, et al., 2001; Woodward, Gosser, and Weiner, 1993) was developed, in part, to address faculty concerns about student learning and high attrition rates in introductory chemistry courses by providing an environment in which students work in problem-solving teams to develop their conceptual understanding and to learn to communicate scientific ideas (Gosser, et al., 1996). (For an overview of the instructional model, see Sarquis, et al., 2001. A comprehensive report of the research and development work on the model is available, see Gosser, et al., 2001.)

The core of the PLTL approach is a weekly 1.5–2 hour peer-led workshop, where students interact to solve carefully structured problems under the guidance of a peer leader. In a typical workshop, six to eight students work as a team to solve carefully structured problems. The Peer leader clarifies goals, ensures that team members engage with the materials and with each other, builds commitment and confidence, and encourages debate and discussion. A good leader keeps the group focused on seeking answers that short circuit thought processes. The peer-leader is central to the model, which is not remedial and is targeted at all students (Sarquis et al., 2001). PLTL is a powerful addition to instructional settings, and is especially useful in settings that do not routinely offer recitation or discussion sessions as a part of their courses. PLTL has also been successfully used to replace recitations led by graduate students (Tien, Roth, and Kampmeier, 2002). Undergraduate peer leaders are students who have been identified as successful in the course, have been out of the course for at least one semester, and who also demonstrate superior interpersonal skills and leadership potential. PLTL has been tested and successfully implemented in chemistry, biology, physics, and mathematics courses at a wide variety of institutions (http://www.pltl.org). In 2002, approximately 15,000 students were enrolled in PLTL courses offered at more than 100 diverse institutions (Varma-Nelson and Gosser, 2004).

In PLTL, the instructors have been routinely responsible for constructing workshop units and for identifying the first cohort of undergraduate peer leaders. Answer keys are not provided to the students or the leaders. The students are expected to develop confidence in the solution through debate, discussion, and persuasion. The emphasis is placed on the process of finding and evaluating several possible answers rather than uncovering the single correct answer. A good PLTL workshop unit should promote brainstorming and teach students to verbally articulate scientific concepts. Ideally, the workshops should end with a summary of the big ideas, with the general accuracy of these conclusions directed by the guiding hand of the peer leader. Reflection by the workshop students on any changes in their own understanding of those concepts is particularly useful, especially when alternative conceptions of the ideas dealt with are common.

Often, PLTL sessions do not involve any graded work or earned credit, so providing a strong and clear added value to the experience is paramount. Without the usual rewards, however, student preparation can be variable. Peer leaders, for example, when surveyed, commonly report lack of student preparation for the workshops (Gafney, 2001). Assigning preliminary readings, appropriate end-of-the-chapter problems, and self-tests are methods used to deal with the problem. Short pre-workshop quizzes are a good way to encourage preparation as well.

Tested workshop materials are available in chemistry (Gosser, Strozak, and Cracolice, 2001; Varma-Nelson and Cracolice, 2001; Kampmeier, Varma-Nelson, and Wedegaertner, 2001). Using existing materials is likely to reduce some initial barriers encountered in implementing the PLTL model because it reduces the work of the initial set-up. Many instructors choose to develop new materials, or modify existing ones, to suit their own student populations and course content.

Sample Workshop Problems

This section presents two problems taken from PLTL Organic Chemistry (Kampmeier, Varma–Nelson, and Wedegaertner, 2001) that work well in a workshop setting and can be used in beginning Organic Chemistry classes. Also included in each case are the instructions given to the student leader.

1. Two isomeric compounds **A** and **B** are known to each have a monosubstituted benzene ring (C_6H_5-). Both have the formula $C_6H_5C_3H_5O_2$, and both are insoluble in water. However, when they are treated with dilute NaOH, **A** dissolves but **B** does not. Give structures for **A** and **B** *consistent* with this information. **Explain your reasoning**.

Discussion: Puzzle problems are great workshop problems. Ask the students to read the problem carefully to find out what they are supposed to do (figure out structures that are consistent with the given observations and the rules of structure). The leader can volunteer to serve as scribe. Ask the students to reread the problem to find the observations. (This is important because everyone can participate.) Now, take the observations one by one, and ask the students to tell what can be deduced about structure from each observation. (Pick respondents carefully, keeping the better students for the tougher observations.) Then, encourage the group to collect the deductions to make larger deductions, and so on, until the problem is solved.

Beginning students tend to try to solve the structure problems all at once, taking all of the data in one big gulp. They also tend to jump to conclusions. They guess a structure on the basis of one observation, for example, $C_5H_{12} = CH_3(CH_2)_3CH_3$, and then they will not let go of it even when the structure is excluded by subsequent observations. It is always much better to have multiple hypotheses (there are three C_5H_{12}'s); then it is psychologically easy to exclude hypotheses.

Take Home Points. Solving problems such as these requires students to know the empirical facts (this structure has these properties or vice versa) and to connect the facts in a purposeful manner. Connecting individual deductions to make larger deductions is both logical and creative. It is distinctly higher-order thinking than knowing the simple observation–deduction pairs.

2. Although we discussed the data of tetrahedral carbon as proposed by van't Hoff and Le Bel, square planar carbon is also a possible structure. Show how this structure can be eliminated by considering CH_2Cl_2 and CH_2BrCl as examples of disubstituted methanes.

a. How many isomers would be present in each instance if the carbon had a square planar structure?

b. How many isomers are possible in each instance if the carbon is tetrahedral?

Consider CHBrClF as an example of a trisubstituted methane.

c. How many isomers would be possible if the carbon atom were square planar?

d. How many isomers are possible for CHBrClF if carbon is tetrahedral?

The experimental observations are:

CH_2Cl_2 one isomer
CH_2BrCl one isomer
CHBrClF   two isomers.

Discussion: This would be a good problem to solve by dividing the group into smaller groups. Students can count isomers and then compare and debate answers.

This is a great workshop problem because it gets directly at the epistemological issues. How is it that we know what we think we know? This problem emphasizes that there is a logical, rigorous way of eliminating the hypothesis that tetravalent carbon has a square planar geometry. The tetrahedral hypothesis is shown to be consistent with observation. The logic does not prove that tetravalent carbon is tetrahedral, but it does prove that it is not square planar.

Take-Home Point. We think what we think because of a trail of logic, not because somebody said it.

Outcomes. The national PLTL Project has engaged in "action research"-based evaluation, in which results have been fed directly back to the programs under study in order to refine and improve the model. Data collection methods have included focus groups, surveys, structured phone interviews, reports from faculty on student grades, site visits, and observations by the PLTL project evaluator Leo Gafney (2001).

Critical components for implementation. This evaluation has identified six critical components for successful integration of the PLTL model into a course. These components have been repeatedly found to contribute to successful student performance, while their absence has led to problems in implementation and lack of gains in student performance and retention. These critical components are as follows:

1. The workshops are integral to the course, not an optional add-on.
2. The workshop materials are challenging, intended to encourage active learning and to work with groups.
3. The workshop leaders are well trained and closely supervised, with attention to knowledge of chemistry and teaching/learning techniques.
4. The faculty teaching the courses are closely involved with the workshops and the workshop leaders.
5. The organizational arrangements (including the size of the group, space, time, noise level, etc.) are structured to promote learning.
6. Workshops are supported by the department and institution.

Exam performance. The PLTL Workshop model has had a positive impact on student attitude and success in the study of science and mathematics. Faculty members have compared the performance of non-workshop and workshop sections of their classes in a given semester or between semesters. Percent success data across a wide range of courses and institutions demonstrate that students who participate in PLTL workshops earn better grades than those who do not (or who may participate in other structures, such as traditional recitations).

One study at Saint Xavier University (SXU) by Varma-Nelson demonstrated that, in a class taken primarily by nursing majors, 99% of whom were women, the student success and retention rate increased by more than 10%

and performance on a national standardized ACS exam improved when the lecture time was reduced from 4 hours to 2.5 hours and the remaining 1.5 hours were used for PLTL workshops. The results on the ACS exam also indicate that content coverage was not compromised.

Tien, Roth, and Kampmeier (2002) conducted a study on groups of students who were in Jack Kampmeier's traditional organic chemistry course from 1992 to 1994 with those who were involved in a PLTL workshop from 1996 to 1999. Although the control and treatment sections were not taught in the same year, they were similar in many ways. The same instructor taught the course; the instructor used the same textbook, lecture style, class size, and level of difficulty. It was found that the workshop participants outperformed the control group on exams in all cases. For overall means, the scores were significantly different with $p < 0.01$. When broken down by gender and ethnicity, the results show that all PLTL groups outperformed their counterparts in the more traditional course. While such rigorous statistical analyses have not been performed in all cases, at least 20 similar studies have been conducted involving PLTL workshops at other institutions involving several disciplines (http://www.pltl.org). As stated by Lyle and Robinson (2003). "Although there may be flaws in a study, if the study is repeated, taking into account the flaws that have arisen and the same general results occur, the results can be considered useful."

Student satisfaction. Surveys completed by more than 700 respondents at nine institutions using PLTL reveal that when the method is introduced with fidelity to the model, students place a high value on the workshops and have found them helpful to their learning. A total of 82% of respondents stated that they would recommend the workshop course to their peers, and the majority felt comfortable asking questions and reported that their leader was well prepared to effectively facilitate the workshops.

Leader reflections. As apparent in the quotes, the leaders assert that their involvement in the PLTL workshops has given them confidence to do more science, improved their presentation skills, and improved their understanding of the teaching and learning process. The following are quotes taken from a study done on former leaders who worked with one of us (PVN) at Saint Xavier University (SXU):

"As a leader, I gained the knowledge and confidence I needed to pursue a career in pharmacy. During many courses in pharmacy school, I became known as a group leader."

"That's always exciting to see—the transition from 'this stuff is just too hard and there's no way I can understand it' and a self-defeatist attitude to these same people explaining the concept to someone else because now they understand it and own it, they teach it now because they've gone through the process."

"This taught me that sometimes students need to see things explained in a different manner from the textbook or lecture notes and handouts…"

"It helped me be more at ease speaking in front of a group, and able to express myself when everyone is looking at me."

"Group Setting: I used to work alone. I never studied with anyone, until workshops. Then I took biochemistry after organic. It was really a tough course. It was the first time it was held.
But I noticed that when I got together with friends, it was as if we were running a workshop. A lot of people don't realize how much it affected them, because they don't think about it. But when you think about changes in the way you learn, you find that you have changed."

"My involvement in the project remains one of the most impressive entries in my resume and has come up in every job interview I've had since leaving St. Xavier U."

For further leader reflections, see essays written by former student leaders from several institutions in the "Peer-Led Team Learning: A Guidebook"(Gosser, et al., 2001).

B. Structured Study Groups (SSG) at the University of Michigan

Introduction. Since 1994, a cohort of 120–160 first-year students earns honors credit in Supplemental Instruction (SI) sessions attached to the 1000-student course of standard coursework and examinations in the organic chemistry based Structure and Reactivity courses (Ege, Coppola, and Lawton, 1997; Coppola, Ege, and Lawton, 1997). We developed this modified supplemental instruction option in lieu of a separate honors section of the course because we felt that first-term students could not judge whether or not an honors section of organic chemistry would be appropriate (Coppola, Daniels, and Pontrello, 2001). In this format, students are able to experiment with taking the course for honors credit for three weeks and then dropping the SSG sessions and taking the course for regular credit without changing their course schedule. We believe that an honors option should broaden a student's experience with respect to the regular course. Because SSG honors students take the same recitation sections, labs, and exams as non-SSG students, coursework for honors and non-honors credits can be compared easily. In addition, an SSG honors option saves a faculty teaching assignment that otherwise would be required for a separate honors section. Finally, the SSG option is not restricted to students in the honors program. Any student who elects and satisfactorily completes the SSG option is awarded honors credit.

SSG leaders are juniors and seniors who demonstrated teaching skills when they were SSG students. Indeed, the SSG assignments permit students to demonstrate their teaching potential with the idea that they might become SSG leaders. SSG leaders identify prospective future leaders from among their students and justify their choices to the faculty coordinator. Identified students submit essays to the coordinator as part of the selection process for SSG leaders. Leaders are not necessarily chemistry majors: of 68 student leaders from 1994 to 2003, 37 have been majors in chemistry, 16 in biochemistry, 4 in biology, 2 in chemical engineering, and 9 in cellular and molecular biology.

As former SSG participants, SSG leaders begin with a strong sense of the program. SSG leaders attend weekly lunchtime meetings with the faculty coordinator to reflect on their teaching, anticipate teaching issues, and determine the evaluation criteria for the week. Each week, a different leader leads a discussion on teaching and learning and then records and distributes the outcomes and recommendations for the group. Leaders continue to discuss teaching issues throughout the week on the SSG leaders' listserv.

SSGs are generally 15–20 students in size. While this is larger than the usual optimum for team size, the SSG sessions structure numerous small group (two to four student) interactions with high accountability, particularly peer review. SSG sessions follow a detailed curriculum that encourages discussion and explanation activities that lead to deep mastery of organic chemistry. In the first session, for example, SSG leaders (1) lead an ice-breaker activity on reciprocal teaching, (2) have students go to the blackboard to teach one another how to decode line formulas, (3) take their group to the library to explore chemistry journals, and (4) present a short introduction to proper citation format. From the first version, created in 1994, all of the instructional materials have been constructed and subsequently modified by one of us (BPC) in full collaborative partnership with undergraduate student leaders.

The first assignment. Beginning with the first homework assignment, SSG leaders have students apply the concepts they are learning to new material using a creative task. In their first assignment, leaders ask that students pick a molecule with 10–13 carbon atoms from a chemistry journal, construct five new (rational) molecules with the same formula, rank the molecules based on selected properties (e.g., magnitude of dipole moment, boiling point, and solubility), and write rationales for their rankings. The student work for this assignment must include a statement putting the journal article into context, a copy of the journal page from which the example came, and a properly formatted citation. In other assignments, leaders ask students to format an appropriate quiz problem from the new material. The second SSG session builds on this homework assignment. Students must submit one copy of their homework to their leader and the other copies are distributed into the group for two rounds of peer review. SSG leaders create an assessment rubric, which, for the first assignment, might address whether the molecules fit the prescribed criteria, whether the format and information are appropriate to the class level, and whether the citation is formatted correctly. Peer review is a time of in-depth discussions and learning; the first round of review can take up to an hour. During this time, the leader circulates, noting common issues that arise, sending students with interesting examples to the board, and otherwise facilitating a usually raucous discussion predicated on one question: *Is this work I am looking at correct or isn't it?* In the peer review, students must grapple with ideas in a classmate's homework that conflict with their own work and, through discussion, must figure out where errors in understanding the material or process have occurred. This grappling gives SSG students the opportunity to make, recognize, and correct their own errors. The reviews are returned to the originator (the review is a piece of paper with "yes" and "no" answers that have been circled), who has a chance to decide whether to make any changes to the original assignment. The SSG leader collects the edited assignments and peer reviews and uses them to evaluate student performance.

One of the overarching goals of the SSG honors sessions is for students to develop the ability to create meaning from new and unfamiliar chemical information, especially scientific information from the primary literature. In the capstone assignment, SSG leaders ask students to read original journal articles, generate and discuss questions that they would ask the author about the article, and meet with the author in a 1–2 hour session to ask their questions. In addition, writing assignments are used (e.g., "create an analogy for chemical phenomenon"), as well as an introduction to research ethics that involves reading and then writing a position paper on a subject of topical interest to science or science education (e.g., the Kansas State Board of Education decision).

The SSG leaders provide feedback to the students on their work and participation in the weekly sessions. They use this feedback to assign grades for the SSG sessions based on a scale of O (outstanding), S (satisfactory), and U (unsatisfactory). Course grades for SSG students are based on a two-part scheme. The base grade for an SSG student is determined in the same manner as it is for a non-SSG student (the exam scores for the course). In order to receive the base grade with an honors designation, students need to receive an "S" to "O" average from their SSG leader. A "U" average results in the student receiving the honors designation, but the base grade is reduced. In the research-oriented section of the course, there are two layers of SSG assignments. For the first layer, SSG leaders help design and then implement a series of tasks that are comparable to those in the SSG sections for the larger course. For the second layer, SSG leaders help students create and carry out projects. In one of these term-long projects, students construct a written and HTML resource on advanced chemical transformations that are incorporated into the course website. Reading journal articles, understanding the chemical transformations described in them, and then creating animated reaction mechanisms and interactive assignments for their classmates cause the students to think like teachers. The multimedia text is fully owned by the students and, as a class-sized group, they must seek out each other's expertise in order to understand the material and complete the project. The SSG leaders in the project-oriented section do a great deal of work that would be impossible for one faculty member to do, including creating lessons on HTML authoring, locating appropriate software, and managing the logistically-demanding task of coordinating the efforts of 100 individuals working on a single website project.

Conclusion

In many respects, team learning is a "second generation" instructional strategy in that it draws together a powerful collection of pedagogies such as *group learning, reciprocal teaching, Vygotsky's learning theory,* and *studio instruction.* Students bring their work on problems to classes structured to promote a reasonable extension of ideas. Theoretically, students are doing generative (studio) work on problems that push them past what they might be able to do individually (their *ZPD*), and they reconvene in groups (group learning) in order to present and aid each other through the navigable steps of the problem-solving process (reciprocal teaching). They engage in such practices as explanation, conversation, peer review, and critique under the watchful eye of a leader.

Learning gains derived from Team Learning include increases in subject matter mastery as well as higher order cognitive skills. In addition, an intimate and trusting social environment creates a network for support within the students' university experiences. Finally, team leaders develop skills in both management and instructional development and implementation.

Suggested Reading:

Boud, D., Cohen, R., Sampson, J. (2001). *Peer Learning in Higher Education.* Sterling, VA: Stylus Publishing.. The first half of the book sets a broad historical context. From the literature, it synthesizes a nice review of designing peer learning environments, describing multiple approaches, discussing management issues, and covering learning and assessment topics. The second half of the book looks at a series of elaborated case studies in MBA and Law School, and then it examines a number of different Instructional Technology examples where distance and other computer-mediated learning are integrated with face-to-face activities. There are a few unique advantages in this particular book. First, the perspective is from adult education rather than elementary education, so the methodologies and discussion can be more easily understood in an advanced context. Second, the editors are knowledgeable and contribute about 50% of the text; their experience as faculty members at the University of Technology, Sydney, is apparent. Third, there is a strong, selective, and not-overwhelming research base to each chapter.

Jaques, D. (2000). *Learning in Groups (3rd ed.)* Sterling, VA: Stylus Publishing. This book is written for people teaching college tutorials in England, but it is an excellent resource for anyone who is involved in higher education. It is easy to read and offers many practical tips on how to use groups in teaching. The book is written in a way that permits flexibility of use. Instead of starting at the beginning, one may start by reading Chapter 6, which provides suggestions for structured activities that can be used in groups. For each activity, there is a discussion on how to set up the task; it also includes benefits and drawbacks. The appendix has a list of problems that one may encounter in group activities, as well as possible solutions. The earlier chapters in the book discuss learning theory, theories of group behavior, and what is known about communication in groups. Chapter 1 has an excellent discussion about how the characteristics of a group change as its size increases. Chapter 9, which is about assessment, is also very useful. There is an extensive list of references for further reading at the end of the book.

References:

Apple, T., and Cutler, A. (1999). "The Rensselaer studio general chemistry course." *Journal of Chemical Education. 76*, 462.

Astin, A. W. (1993). *What Matters in College?* San Francisco: Jossey-Bass. . 394.

Astin, A. W., and Astin, H. S. (1993). *Undergraduate Science Education: The Impact of Different College Environments on the Educational Pipeline in the Sciences.* Los Angeles, CA: Higher Educational Research Institute, UCLA. ERIC Report, NSF Grant SPA-8955365.

Atkinson, J. (2001). *Developing Teams Through Project-Based Learning* Hampshire, UK: Gower Pub.

Bailey, C. A., Kulinowski, K., and Paradis, J. (1998). "Studio chemistry: A feasible environment for large general chemistry courses?" Abstracts of Papers of the American Chemical Society 215, 091.

Blunt, J. A., Sweeder, R. D., Bartolin, J. M., Gottfried, A. C., Coppola, B. P., and Banaszak Holl, M. M. (2003). "Preparing future faculty for students at all levels of the education process: Studio 130," Abstracts of Papers of the American Chemical Society. 225, 1248.

Boud, D., Cohen, R., and Sampson, J., (Ed.) (2001). *Peer Learning in Higher Education: Learning From and With Each Other.* London, UK: Kogan Page.

Bransford, J. D. and Schwartz, D. (1999). "Rethinking transfer: a simple proposal with multiple implications." In *Review of Research in Education.* Washington, DC: American Education Research Association.

Brown, A. L., and Palincsar, A. S. (1989). "Guided, Cooperative Learning and Individual Knowledge Acquisition." In L. B. Resnick (Ed.), *Knowing, Learning, and Instruction: Essays in Honor of Robert Glaser.* Hillsdale, NJ: Lawrence Erlbaum Associates. 393–451.

Bruffee, K. A. (1995). "Sharing Our Toys" *Change. 27* (1), 12–18.

Chambers, B., and Abrami, P. C. (1991). "The relationship between Student Team Learning outcomes and achievement, causal attributions, and affect." *Journal of Educational Psychology*. 83, 140–146.

Coleman, E. B., Brown, A. L., and Rivkin, I. (1997). "The effect of instructional explanations on learning from scientific texts." *Journal of the Learning Sciences.* 6(4), 347–365.

Coleman, E. B. (1998). "Using Explanatory Knowledge During Collaborative Problem Solving in Science." *Journal of the Learning Sciences.* 7(3–4), 387–427.

Coll, R. K., and Taylor, T. G. N. (2001). "Using constructivism to inform tertiary pedagogy." *Chemistry Education: Research and Practice in Europe.* 2(3), 215–226.

Coppola, B. P. (1996)."Exploring the Cooperative and Collaborative Dimensions of Group Learning." *Chemical Educator.* 1(1): S 1430-4171(96) 01006-0.

Coppola, B. P. (May 5, 2004). This supposition was supported by Michaelsen during the 2004 SUN Conference at the University of Texas, El Paso, where one of the authors (BPC) and Michaelsen were both keynote speakers; private communication from LKM to BPC. Available online at *http://www.utep.edu/cetal/sun/2004/.*

Coppola, B. P., Daniels, D. S., and Pontrello, J. K. (2001). "Using Structured Study Groups to Create Chemistry Honors Sections." In J. Miller, J. E. Groccia, D. DiBiasio (Eds.) *"Student Assisted Teaching and Learning."* New York: Anker.116–122.

Coppola, B. P., Ege, S. N., and Lawton, R. G. (1997). "The University of Michigan Undergraduate Chemistry Curriculum. 2. Instructional Strategies and Assessment." *Journal of Chemical Education.* 74, 84–94.

Coppola, B. P., and Pearson, W. H. (1998). "Heretical Thoughts II. These on Lessons We Learned from our Graduate Advisor that have Impacted on our Undergraduate Teaching." *Journal of College Science Teaching.*27, 416–421.

Ege, S. N., Coppola, B. P., and Lawton, R. G. (1997). "The University of Michigan Undergraduate Chemistry Curriculum. 1. Philosophy, Curriculum, and the Nature of Change." *Journal of Chemical Education.* 74, 74–83.

Gafney, L. (2001). "Workshop evaluation." In D.K. Gosser, M.S. Cracolice, J.A. Kampmeier, V. Roth, V.S. Strozak, and P. Varma-Nelson (Eds.) *Peer-Led Team Learning: A Guidebook.* Upper Saddle River, NJ: Prentice Hall.

Gafney, L and Varma-Nelson, P. (2002). "What happens next? A follow-up study of workshop leaders at St. Xavier University." *The Workshop Project Newsletter.* Progressions: Peer-Led Team Learning. 3(2):1, 8–9

Gosser, D.K., Cracolice, M. S., Kampmeier, J. A., Roth, V., Strozak, V. S., and Varma-Nelson, P. (Eds.) (2001). *Peer-Led Team Learning: A Guidebook.* Upper Saddle River, NJ: Prentice Hall.

Gosser, D. K., Roth, V., Gafney, L., Kampmeier, J. A., Strozak, V., Varma-Nelson, P., Radel S., and Weiner, M. (1996). "Workshop chemistry: Overcoming the barriers to student success." *The Chemical Educator* [On-line serial]. 1, 1–17.

Gosser, D. K., Strozak, V. and Cracolice, M. (2001). *Peer-Led Team Learning: General Chemistry.* Upper Saddle River, NJ: Prentice Hall.

Gottfried, A. C., Hessler, J. A., Sweeder, R. D., Bartolin, J. M., Coppola, B. P., Banaszak Holl, M. M., Reynolds, B. P., and Stewart, I. C. (2003). Studio 130: Design, testing, and implementation. *Abstracts of Papers of the American Chemical Society.* 225, 647.

Hills, H. (2001) *Team-Based Learning.* Hampshire, UK: Gower.

Hoffmann, R. (October 5, 2003). Private communication to BPC. "The research group social/family structure is one of the great inventions of American graduate education, a part of what makes our graduate students catch up to their Asian and European counterparts, when they start out two years behind (my guess) at the beginning of graduate school. The group is where two other essential pieces of education take place—the moral education, for better or worse, seeing what the professor thinks is ethically good, what he or she allows to slip by. It is also where one learns what the unspecified quality judgments of science (are), what work is routine, what is more than that."

Johnson, D. W., and Johnson, R. T. (1999). *Learning Together and Alone.* Englewood, NJ: Prentice Hall, Inc.

Kampmeier J. A., Varma-Nelson, P., and Wedegaertner, D. (2001). *Peer-Led Team Learning: Organic Chemistry.* Upper Saddle River, NJ: Prentice Hall.

Katzenbach, J and Smith, D. (1993). *The Discipline of Teams.* Harvard Business Review.

Lencioni, P. M. (2002). *The Five Dysfunctions of a Team: A Leadership Fable.* San Francisco: Jossey-Bass.

Michaelsen, L.K. (2004). *Getting Started with Team Learning.* Available at *http://www.ou.edu/idp/teamlearning/materials.htm.*

Michaelsen, L. K, Knight, A.B., and Fink, L. D. (2002). *Team-Based Learning: A Transformative Use of Small Groups.* Westport, CT: Praeger Publishers.

NSF: The National Science Foundation VIGRE (Vertical Integration of Research and Education) programs bring together faculty, postdoctoral fellows, graduate students, and undergraduate students at all levels of research and education to discover, inform, and teach, in other words, to facilitate formation of research teams. Available online at *http://www.nsf.gov.*

Palincsar, A. S., and Brown, A. L. (1984). "Reciprocal Teaching of Comprehension-fostering and Comprehension-monitoring Activities." *Cognition and Instruction.* 1, 117–175.

PLTL website. Available online at *http://www.pltl.org.*

Rieber, L. P. (2000). "The Studio Experience: Educational reform in instructional technology." In Brown, D.G. (Ed.) *Best Practices in Computer Enhanced Teaching and Learning.* Winston–Salem, NC: Wake Forest Press.

Sarquis, J. L., Dixon, L. J., Gosser, D. K., Kampmeier, J. A., Roth, V., Strozak, V. S., and Varma-Nelson, P. (2001). "The Workshop project: Peer-led team learning in chemistry." In, J. E. Miller, J. E. Groccia, and M. Miller (Eds.), *Student-Assisted Teaching: A Guide to Faculty-Student Teamwork* Bolton, MA: Anker Publishing Company 150–155.

Schön, D. A. (1983). The Reflective Practitioner: How Professionals Think in Action. New York: Basic Books.

Seymour, E., and Hewitt, N. (1997). *Talking About Leaving: Why Undergraduates Leave the Sciences.* Boulder, CO: Westview.

Steele, C. "Thin Ice: 'Stereotype Threat' and Black College Students". In, N. Davis and R. Robinson (Eds.) *Sociological Perspectives on American Society, 4th ed.* New York: Simon and Schuster (Pearson Press).

Sweeder, R. D., Bartolin, J. M., Hessler, J. A., Gottfried, A. C.,. Coppola, B. P, Banaszak Holl, M. M., McKeachie, W. J., and Stewart, J. R. "Studio 130: Evaluation." (2003). *Abstracts of Papers of the American Chemical Society.* 225, 1262.

Tenney, A., and Houck, B. (2004). "Learning about Leadership: Team Learning's Effect on Peer Leaders" *Journal of College Science Teaching.* 33, 25–29.

Tien, L. T., Roth, V., and Kampmeier, J. A. (2002). "Implementation of a Peer-led team Learning Instructional Approach in an Undergraduate Organic Chemistry Course." *Journal for Research in Science Teaching.*, 606.

Torres, C., and Spiegel, J. (1991). *Self-Directed Work Teams: A Primer*. San Francisco: Jossey-Bass.

Triesman, U. (1992). "Studying students studying calculus: A look at the lives of minority mathematics students in college." *The College Mathematics Journal.*23 (5).

Varma-Nelson, P., and Gosser, D. K. (2004). "Dissemination of Peer-Led Team Learning and Formation of a National Network Embracing a Common Pedagogy." In *Teaching Inclusively: Diversity and Faculty Development*. M. Ouellett (Ed.) New Forums Press. (accepted for publication).

Varma-Nelson, P., and Cracolice, M. (2001). *Peer-Led Team Learning: General, Organic, and Biological Chemistry*. Upper Saddle River, NJ: Prentice Hall.

Vygotsky, L. (1978). *Mind in Society: The Development of Higher Psychological Processes*. Cambridge, MA: Harvard University Press.

Vygotsky, L. S. (1985). *Thought and Language* Cambridge, MA: The M.I.T. Press.

Wilson, J. (1994). "The CUPLE Physics Studio." *The Physics Teacher*. 32, 518.

Woodward, A., Gosser, D. K., and Weiner, M. (1993). "Problem-Solving Workshops in General Chemistry." *Journal of Chemical Education*. 70, 651–665.

Part

III

LEARNING WITH TECHNOLOGY

Electronic Data Collection to Promote Effective Learning During Laboratory Activities

Norbert J. Pienta
Department of Chemistry
University of Iowa

John R. Amend
Department of Chemistry and Biochemistry
Montana State University

Abstract

The focus of undergraduate teaching laboratories may be on the teaching or learning of concepts, but these goals are often complicated by a desire to provide students with a tactile or "hands-on" experience. As a result, experiments that may have been designed to enable the students to develop and confirm a principle have often evolved into "cookbook" recipes that require little or no thinking. That the laboratory can offer the opportunity to learn problem solving and the scientific method is apparent from a successful apprenticeship model that is applied as graduate students learn to conduct independent research. The larger number of students in the undergraduate enterprise and their limited knowledge of chemistry are the basis of the challenge. Indeed, several decades of research on learning from the undergraduate laboratory experience using "traditional" experiments do not even provide clear evidence of learning. However, several positive changes have appeared in the last decade. The implementation of good pedagogical practices for the laboratory include guided inquiry, cooperative learning, and the science writing heuristic, for example. Furthermore, these can be coupled with technology in the form of computers and interface devices that enable the collection, visualization, and manipulation of research quality data, and they give the best hope of observing and measuring effective teaching and learning in the laboratory.

Biographies

Norb Pienta has been an associate professor and director of undergraduate studies at the University of Iowa since 1999. He has recently implemented major changes in the general chemistry curriculum, including the reintegration of the laboratory with lectures and changes that include a more active student role and a cooperative learning environment in discussion and laboratory sessions. Previously, he spent a decade as a director of laboratories at the University of North Carolina at Chapel Hill. In that role, he developed new experiments and curricula for all the chemistry laboratory courses, including student data collection using personal laptops.

John Amend is a professor at Montana State University. His research interests are in analytical chemistry and chemical education, including the development of computer-based instrumentation for research and teaching and the development of interactive computer models of natural resource systems. John has been developing hardware and software for laboratory data collection and using them in general chemistry laboratories since the 1970s.

The focus of undergraduate teaching laboratories

Anyone who teaches in an undergraduate laboratory program may wonder why chemistry departments use a different, fundamental model for the preparation of undergraduate than for graduate students. The apprenticeship or "research" model involves contemporary problems about the nature of chemistry conducted using state-of-the-art approaches and instrumentation following the scientific method. However, at some of the best "research" schools (and other exclusive ones known for their "teaching"), it would not be unusual to find undergraduate students in introductory courses following a well prescribed "cookbook" recipe or verification approach to re-create an observation or confirm a principle that has appeared in texts for decades or even a century. A recent survey suggests that over 90% of general chemistry laboratories employ this plan (Abraham et al., 1997).

The laboratory content or approach to teaching may not be the only problem. Particularly in times of financial exigency, chemistry departments are reminded that the laboratories require extensive resources in space, money, and time. Some laboratory courses are essentially neglected; the curriculum, indeed the syllabus, may be many years old. At larger universities, the curriculum and teaching methods used in the laboratories are often the exclusive territory of lecturers and staff; tenured faculty members may find that conducting a modern, pedagogically-sound laboratory curriculum may be too hard, require too much of their time, or need the continuity that personnel "devoted" to the task may require.

Although it may be a challenge to design effective laboratory instruction, strategies have begun to emerge, particularly since the early 1990s. Examples of effective pedagogical changes coupled with benefits that accompany modern instrumentation, technology, and computers are being documented. This chapter starts with reviews of research studies from the last several decades.

A historical perspective on learning in the laboratory

Bates (1978) looked at 79 research studies, primarily from secondary education. He concluded that most research failed to assess outcomes specific to the laboratory, although it appeared that lecture, demonstrations, and laboratory work are equally effective. He suggests that some inquiry-based laboratory activities are well suited to teaching the process of inquiry with the caveat that teachers need to be skilled in those teaching methods. Furthermore, there is great potential for promoting positive student attitudes and providing a large variety of different learning opportunities. The review is positive toward the value of the laboratory experience but is the source of the quote that may reappear in the budget-limited 2000s: "What does the laboratory accomplish that could not be accomplished as well by less expensive and less time-consuming alternatives? (Bates 1978, p. 75)."

Blosser (1983) observed that much of the literature was based on opinions and then analyzed previous studies (at the primary, secondary, and college level) that could be evaluated using statistical analysis techniques. She reported 191 treatments of dependent variables that could be clustered into groups: achievement, attitudes, cognitive abilities (including reasoning and critical thinking), the process of science, understanding of science, manipulative skills, student interests, dogmatism, retention, and ability to engage in independent work. Of the 191 quantitative treatments, 29 supported laboratory as a positive feature at a statistically significant level, 7 favored other features, 16 contained mixed results, and 139 showed no statistical significance.

Given the expectation that laboratories in sciences mimic direct experiences with the physical world, it is remarkable how many studies are inconclusive. Hofstein and Lunetta (1982) suggested that many studies failed because they suffered from several weaknesses, including: the selection and control of variables, the group size, the effect of teaching, the laboratory manual, and complications from the use of instrumentation. In other words, differences among evaluated groups were hidden by confounding variables for which there were no controls, evaluative instruments that were not sensitive to the variables and poor experimental design. These factors and implications for research from the past (Nakhleh, Polles, and Malina, 2002), however, can serve as guidance for future chemical education research.

One objective is to decide whether the laboratory has value or whether it is an outdated artifact that should be abandoned. Two proponents of the latter view are Hodson (1990), who opined that experiments are usually poorly conceived, confusing, and unproductive, and Toothacker (1983), who felt that because introductory physics labs had not worked, laboratories should be eliminated. Kirschner and Meester (1988) suggest that simple verification of principles already known to students reduce student motivation and curiosity and thus are a poor return on an expensive investment. Furthermore, trivial experiments are a waste of time and perceived by students as such; non-trivial ones are often beyond the students' understanding and do little to advance it. In spite of these negative impressions, more recent studies demonstrate that laboratory instruction has value; they are summarized in the following sections.

Positive results in recent times

Effective pedagogy in systematic changes. Guided inquiry, cooperative learning, the science writing heuristic, learning cycles, and visualization are the bases of other contributions to this volume and should play an integral role in the laboratory experience. For example, Cooper (1994, 1995) reports on the successful use of cooperative learning techniques (including group work, group grades, peer evaluation, and inquiry-based experiments) coupled with multimedia software that provides some of the context. She reports positive attitudes, higher letter grades, and lower drop rates, particularly among women when compared to traditional laboratories. Rudd, Greenbowe, and Hand (2001, 2002) have used guided inquiry coupled with the science writing heuristic in general chemistry at a midwestern university and have achieved positive results with student performance on lecture exams, laboratory practical exams, and ACS standardized exams. Tien, Rickey, and Stacey (1999) and Rickey and Stacey (2000) demonstrate the benefits of group work, cooperative learning, and the science writing heuristic. Ricci et al. (1994) report a curriculum for the first four semesters of college chemistry (i.e., general and organic chemistry) in which discovery laboratory experiments provide the basis for the learning of concepts during subsequent lectures. Bailey et al. (2000) have developed a truly integrated lecture–laboratory environment in a studio classroom where each of the 64 student stations serves as their common work area for all components of the course. Computers are used for data acquisition, molecular modeling, tutorials, and simulations. Data from the first three years were limited to positive student responses to the facility and curriculum; faculty and visitors observed increased student involvement compared with the traditional mode. Additional examples of the effective use of learning theories also appear in subsequent sections.

Visualization. Early outcomes from microcomputer-based laboratories (MBL) suggested them as a powerful means for communicating data. Thus, MBL represented data in several ways, graphed it in real time, and allowed students to relate the physical event to data in the graph. Furthermore, it allowed them to concentrate on the interpretation, rather than the creation, of the graphed information (Mokros and Tinker, 1987). Nakhleh (1994) reviewed over 20 reports on microcomputer-based labs and how they have affected science learning. That review organizes the results according to type of probe (i.e., data collected: heat and temperature, motion, and pH) but also summarizes studies about student learning from graphs. More recent work appears in a review by Lapp and Cyrus (2000) that focuses on the benefits of "graphing technology coupled with data-collection devices" to mathematics and science classrooms. They conclude that, although the literature suggests that studying graphs can lead to a deeper understanding of physical concepts, students nonetheless have difficulties with connecting graphs with physical concepts and the real world and with transitioning between graphs and physical events. Even with more abstract concepts in chemistry like acidity and basicity, Nakhleh and Krajcik (1994) speculate that on-screen graphs allow students to focus more on what is happening in an activity without relying on short-term memory for the actual data that they collected.

Student attitudes. There are many suggestions that the value of laboratories might not be measurable in a quantitative sense or via traditional tests of concept learning but are more likely to be found in the affective domain, in the form of student motivation and curiosity that carries their interest throughout the course (Roth, 1994; Pickering, 1993; Nakhleh et al., 2002). The laboratory portion of a course may provide the motivation that serves the students throughout the entire course.

Goals for teaching laboratories

One can postulate about how the popularity of some of the current laboratory practices evolved, including traditional "cookbook" experiments. The desire to teach *both* concepts *and* techniques (i.e., laboratory skills of a chemist or a "vo-tech" approach) is likely to create a mixed or confused course objective. Is there a set of goals that are generally agreed upon, and can a small collection of these be reasonably accomplished during a laboratory meeting? Kirschner and Meester (1988) suggest that, in spite of wide agreement on the value of the laboratory experience, often goals are too specific or general to be widely useful and have catalogued over 120 objectives! They argue that labs have far more potential than simply illustrating lecture ideas or measurement techniques. Abraham and coworkers (1997) asked coordinators of the first general chemistry course at colleges and universities to rank five laboratory goals. Based on survey respondent rankings (1 = highest to 5 = lowest), the averages are 2.12, 2.43, 2.49, 3.71, and 4.31 for the following goals: concepts, laboratory skills, scientific processes, positive attitudes, and learning facts, respectively. In their discussion, they note that these data are at odds with students' perception that factual knowledge is more heavily stressed.

The needs are obvious—bridge the gap between the instructors' goals and the students' perceptions and do this in an interesting and pedagogically sound way. What suggestions are there on how to do this? Nakhleh, Polles, and Malina (2002) point out that the laboratory is a complex environment in which students interact with each other, with the laboratory activity, with the equipment or instruments, and with the instructor(s). Those interactions include cognitive, affective, and psychomotor realms. Based on their review, Nakhleh et al. (2002) provide implications for teaching that promote the implementation of microcomputer-based laboratories (MLB):

- practical, real-world connections using up-to-date technology
- pre- and post-laboratory oral discussions
- limited, specific goals
- procedural skills and instrument use that are clustered (i.e., using skills from the first experiment in the second and so forth)
- ability to allow students to ask, "What if?"

Lazarowitz and Tamir (1994) analyzed 37 research reviews covering 1954 to 1990 and distilled their findings into four goals that are consistent with college chemistry laboratory education:

- accomplishing [data collection and] manipulation via computers and technology
- developing student skills in logical thinking and in organization
- providing concrete or practical examples, including ones that confront misconceptions
- constructing a system of values related to the nature of science.

Promising changes

Advantages of electronic data collection. Computing hardware and software is quite pervasive, having moved from desktop computers to compact laptops, tablet PCs, pocket-sized handheld devices (PDAs) and beyond to very powerful cellular telephones. The increase in calculational power and decreasing price means that electronic data collection (EDC) devices can be coupled to a host of sensors and thereby reproduce the function of many laboratory instruments. This suggests that even introductory laboratory students potentially have access to a variety of research quality data. Laboratory experiments conducted using EDC devices promote the goals suggested in the previous section, particularly when coupled with effective pedagogy. The perceived advantages of modern data collection systems entail

- devices that collect research quality data from a wide variety of sensors (ca. 80 sensors are commercially available)
- software that is easy to use, flexible, and contains several modes for data manipulation and visualization
- experiments that can be designed to solve "real" (i.e., practical) problems that are interesting and empowering to students
- efficiency of time utilized during data collection that enables students to explore the experimental variables and to do repeated experiments in a limited laboratory period.

The microcomputer-based laboratory (MBL) or calculator-based laboratory (CBL) continues to evolve, so we opted to use the more general "electronic data collection" in the title of this chapter. We will provide details about the use of EDC devices and the learning opportunities they afford after a brief review about the instrumentation and its application to data collection in teaching laboratories.

Instrumentation and data collection.   Microcomputers appeared in physics and chemistry labs in the late 1970s, and mass production of inexpensive interface devices was introduced in the early 1980s (Vendors, 2003). Generally, early models were expensive and only emulated a single instrument; current versions sell for less than $1000 per student station and are capable of interfacing over 80 sensors or probes with a computer, calculator, or PDA (handheld computer). Sensors include ones to detect light (intensity or color), heat, motion, force, pressure, the presence of a host of solutes in water (e.g., pH or acidity, ions, and gases), and many others.

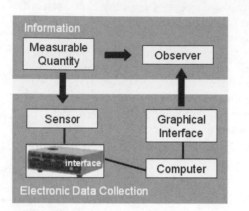

A schematic of an electronic data collection (EDC) system is shown in Figure 1. The top block in the figure represents the expected information transfer—a measurable quantity or property that should be provided to an observer. The EDC system comprises a sensor (capable of measuring the quantity or property), an electronic interface (that transfers the data from the sensor), a computer (that controls the interface and manipulates data), and a graphical interface (a means for viewing the data in a graphical form). If the sensor is a pH electrode, the EDC system acts like a pH meter; if the sensor is a thermistor, the EDC becomes a calorimeter. A simple change in one of many inexpensive sensors produces new instrumentation.

Figure 1. A schematic of an electronic data collection system.

Starting in the early 1980s and continuing to the present, the chemical education literature has evolved as it follows the development of interfaces, sensors, and experiments, or even curricula that espouse their use. For example, circuitry that served as an interface device (like one to conduct electrochemical experiments) were reported early (Grimsrud and Amend, 1979). As small instruments and devices came equipped with connectivity to computers (like serial ports), a series of reports provided technical details and often the context or experimental setting in which they could be used: multimeters (Viswanathan, Lisensky, and Dobson, 1996) or spectrophotometers (Amend, Morgan, and Whitla, 2000).

In the mid-1990s, calculator-based laboratories gained in popularity, particularly in secondary schools. The major emphases have been (1) student graphing skills from a mathematical perspective (Caniglia, 1997; Nicol, 1997; Lapp, 2000), (2) physics and motion (Brueningsen and Bower, 1995; Nemirovsky, Tierney, and Wright, 1998), and (3) chemistry (Donato, 1999; Roser and McCluskey, 1999; Sales, Ragan, and Murphy, 1999; Hickman, Helburn, and Delinger 2000; Cortes–Figueroa and Moore, 2002).

During this time, the MBL focus turned to the experiments and to the manipulation of the data. Thus, with commercial interfaces and sensors, scientists could turn their attention to developing experiments and demonstrations based on more complex systems like photosynthesis (Choi et al., 2002) or building a glucose biosensor (Choi and Wong, 2002). Vitz (2002) developed an elegant yet simple system, LIMSport, over a 15-year period. Thus, he built a hardware and software system that would automatically acquire data directly from its source into a spreadsheet. Students learn spreadsheet techniques for data analysis and presentation, a set of skills that are generally useful and "exportable." Indeed, one of Nakhleh's "implications for teaching" (Nakhleh, Polles, and Malina, 2002) is the suggestion to cluster or re-use skills. A multi-functional spreadsheet that mimics this component of a commercial office suite of software is common not only to LIMSport but to the software in all other commercial data collection packages.

Up to the introduction of LIMSport, the examples cited were individual examples of applications or experiments. As the data collection devices (both hardware and software) became more powerful *and* educators saw a series of potential applications, some individuals began to integrate effective pedagogy and to create entire curricula. Examples of systemic changes were already cited in the previous "effective pedagogy" section (Ricci et al., 1994; Bailey et al., 2000), and this trend seems to be expanding (Durick, 2001; Bondeson, Brummer, and Wright, 2001; Pienta et al., 2003).

Abraham (1997) found that, at the time, only 26% of colleges used computers in general chemistry labs and only 12% used them for two or more experiments. (With the exponential growth of computers reported since that time, these numbers are likely to be much larger.) Both authors of this chapter have adopted effective pedagogy with electronic data collection in large enrollment general chemistry courses. One of us has had such a program for over 20 years. The following sections outline the effectiveness of our laboratory models.

Emulating the research model. Earlier in this chapter, we posed the question about why the research model was not being used extensively in teaching laboratories. Figure 2 shows a schematic representation of the process of problem solving. In the top block of Figure 2, the apparent task for the student is to find a solution to the problem. The steps are the ones of the scientific method, but they immediately point out why this model likely was abandoned. In traditional verification experiments, the student activities are shown in the lower block—data collection, processing, and organization of the data, reflection on the results, and conclusions that provide the solution. Experimental design has been eliminated, as has the opportunity or need to repeat the activities in the lower block. Time is the likely culprit. It takes a considerable portion of the two- or three-hour laboratory period to collect the data necessary to answer the contrived question. In order to make the

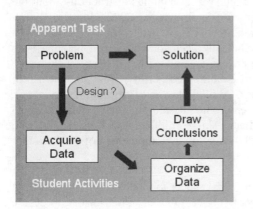

experiment possible, the instructor (1) defined the problem so that no time was lost on unfocused work, (2) described the procedure explicitly to guarantee success under conditions of efficiency, and (3) assessed the students on their ability to get the "right" answer. The instructor and teaching assistants work with the students to avoid pitfalls. The students follow the instructions as carefully as possible. They take their data home to perform the calculations or to analyze it. There is no need for discussion or for any lengthy reflection since the outcome of the experiment is never in question. Although verification experiments are well intentioned, student interest and learning is minimal (Nakhleh, Polles, and Malina, 2002).

Figure 2. Schematic of the problem-solving process during a traditional laboratory session.

EDC devices provide several advantages both with respect to providing additional time for a better learning process and in terms of the quality of the data collected. Although some might argue that a data collection device is a "black box," such a system frees students from the monotony and questionable value of recording

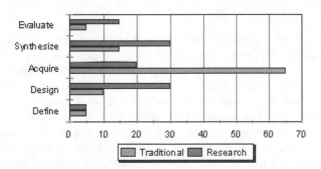

data. In virtually all instances, there is time to repeat experiments. This supports learning about the process of science in which understanding of a concept is built through examination of some of the variables. There is clear evidence that repeated student activities with devices can improve students' understanding about physical phenomena (Brasell, 1987; Nemirovsky, Tierney, and Wright, 1998; Lapp and Cyrus, 2000).

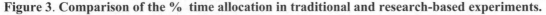

Figure 3. Comparison of the % time allocation in traditional and research-based experiments.

Figure 3 shows a comparison of work and time allocation in traditional verification experiments versus ones conducted like a research problem: (1) to *define* the original problem requires professional maturity and remains the responsibility of the course instructor under either model; (2) the *design* of the initial experiment need only be sufficient to start the cycle and provide some guidance; (3) to *acquire* data requires the least insight and experience and provides the largest time savings with interface devices; (4) the student can *synthesize* the concept after he/she organizes and analyzes the data; in the research model, this step is done in the laboratory, not at home, and is critical to the learning process; and (5) the student must *evaluate* his/her results, discuss them, and improve on the experimental design thereby starting the cycle of these steps again. The student has the opportunity and the need to invoke a new hypothesis and design an additional experiment to test the variables.

A cycle of discovery and visualization. A substantial amount of research has been directed at connecting graphs with the events that generate the data and the physical concepts that dictate the behavior (Mokros and Tinker, 1987; Nakhleh and Krajcik, 1994; Nemirovsky, Tierney, and Wright, 1998; Lapp and Cyrus, 2000). The suggestion is that on-screen graphs allow students to focus more on what is happening in an activity. All of the commercial electronic data collection systems comprise software that is able to store and exhibit data in the form of spreadsheets and to plot them graphically in addition to being able to control the interface device. One of the criteria for selecting EDC systems ought to be the ease with which one can manipulate the data "on the screen." For example, if the headings can be "dragged-and-dropped" from the spreadsheet to an axis of a plot, students will be more likely to examine a mathematical relationship graphically. In other words, if the plot of pressure versus volume is not linear, is the plot of pressure versus 1/V linear?

Students become very comfortable using EDC systems very quickly, especially if they receive orientation to the hardware and software in a first experiment and if a "measurement manual" that contains instructions about the hardware and software is available (Fursteneau, 1990; Morgan, 1997; Rigeman, 2000; Meade, 2002). Typically, greater than 90% of the class expresses satisfaction with the technology. After just a short period of observing students using EDC systems, it is quite clear that they become accustomed to, and perhaps even dependent on, the visual representation of their data. (That they are understanding the concepts behind the experiment and the real meaning of the graphs is another issue and is addressed below.) The student use of multiple graphical interfaces on a single screen (i.e., spreadsheet, X–Y plot, flowchart of the data collection steps, and the current data point) provides a powerful learning environment, although how well students assimilate the variety of information "streams" remains to be studied. This would suggest a disadvantage to computing devices with very small screens (e.g., calculators and PDAs), although some users of PDAs might suggest otherwise (Reeves and Ward, 2003; Pienta and Tardy, 2004).

Data collection software can provide several levels of sophistication with respect to a user's knowledge about the data collection process. Commercial EDC devices allow the user to connect one or more sensors, to decide when and how often to collect data, and to decide whether to "assemble" the data in a spreadsheet and/or on a graph. For the novice user, it may not be necessary (or useful) for the student to see the actual steps of a data collection protocol. It may be sufficient for those students to know that a light intensity datum arrives every few seconds. Once the students become more experienced and sophisticated, they will want to know more. Some of the software systems allow the user to design the actual data collection protocol or series of steps that the computer will follow. This feature, which is implemented via drag-and-drop instructions or steps, enables the user to design the experiment or at least to understand the data collection process at a higher level.

A complete problem-solving process. EDC systems have demonstrated their value in reducing the data acquisition time, in collecting research quality data, in permitting visualization of the experiment and data collected, and in promoting experimental design. It is now time to examine the laboratory problem-solving cycle taking these features into account. Several studies have appeared as a result of a general chemistry laboratory curriculum at a western state university that has promoted the use of data collection devices (Fursteneau, 1990; Amend et al., 1990; Amend, Tucker, and Fursteneau, 1990, 1991; Amend and Fursteneau, 1992; Morgan, 1997; Amend, Morgan, and Whitla, 2000). Over a 20-year period, the laboratory model has evolved to include discovery-based experiments on practical problems, manuals for instrumentation and software use, device-control software that allows student design of the experiment and data collection, and a cooperative learning environment involving group work in a laboratory facility recently remodeled to expedite the curricular features.

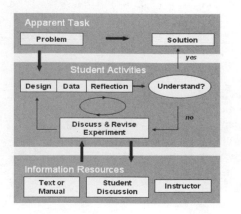

Figure 4. A scheme for a complete problem-solving process.

Figure 4 extends the model introduced in Figure 2. Once again, the top block defines the apparent task and the one below it contains the student activities. A problem is posed to the student. He/she must design a first experiment, acquire the data, organize and analyze it, and reflect on the significance of the outcomes. Unlike the previous model, the complete scheme now enables or requires an evaluation step. Does the student have enough data and understand it well enough to be able to supply a solution? The expectation is that he/she would not in the first cycle. At this

stage, it is appropriate to discuss and revise the experiment. This can involve interactions with fellow students, with an instructor, and with the laboratory activity (in the form of a text, laboratory manual, website, or some other resource). That leads to the formulation of another plan or design. The design, data, and reflection components are repeated. In fact, this process can and should proceed until the students have examined several variables. Much or most of the laboratory period is used in this iterative process. Various features of the EDC system expedite the process: the actual performance of each experiment is quick. The ability to visualize the data enables the user to make a judgment about its utility and value toward a solution. The data is of research quality so that small changes are apparent and the process is very reproducible. The students can ask themselves what a plot should look like and whether that is what they are observing. At one of our institutions, the computers are networked in a way that permits selection and projection of any student's data. Results that are good, bad, or just interesting are generally a springboard for spontaneous discussion.

Figure 4 is a better match to the research model used in graduate teaching in chemistry. The problem can be one that is more open-ended and although it may be contrived, at least it will have the appearance of being important and significant. There is not one path to solving the problem nor is there a single order to examining the variables. Electronic data collection promotes discovery-based learning and the use of inquiry takes advantage of EDC system features.

Besides being compatible with guided inquiry, the EDC systems can be used with a variety of other effective pedagogies. It is often convenient for students to work in pairs because of available resources. Cooperative and group work fit the model and provide a mechanism for the part of the cycle relating to the discussion and revision. Laboratory reports based on the science writing heuristic (SWH) would offer several advantages over traditional ones. The EDC approach is consistent with and synergistic to SWH. Both involve processes that are formative—understanding is built through a series of stepwise activities. Writing complements the discussions that are promoted throughout the experimental process.

Student attitudes about electronic data collection. The last several sections outline the advantages of EDC and provide some models to understanding their effectiveness in teaching and learning. It has been suggested (Nakhleh, Polles, and Malina, 2002; Roth, 1994; Pickering, 1993) that the value of laboratories might not be measurable in the quantitative sense, but may appear in the affective domain in the form of attitude or motivation to persevere in constructing understanding. Fursteneau (1990) reports student opinion data: over 75% of the students felt that doing MBL experiments was an excellent or good idea and only 8% thought that it was a bad or very bad idea; however, 9% strongly agreed and 67% agreed that those labs were challenging. The popularity of MBL was further evidenced in answers to the question of whether computers are useful for learning chemistry in the laboratory. In the role of designing experiments and acquiring and analyzing data, 49% of the students found computers very useful and an additional 30% found them useful. A manual to explain the use of the hardware and software has been perceived as critical to students' facile use of them; 11% found this manual difficult or very difficult, 31% were neutral on the subject, and the remaining 58% found it easy or very easy to understand and use. Morgan (1997) found that student attitudes about a discovery experiment were positive for student enjoyment (38% like the discovery experiment better than the traditional, 51% liked them to the same extent) and for perceived student learning (34% claimed they learned more in the discovery versions, 50% learned about the same amount).

A scholarly work is not supposed to contain the words "fun" or "cool," but data collection systems match students' general interest in electronic devices. The most popular devices are those that provide access to a large amount of information (i.e., a huge bandwidth) while simultaneously providing the user with all or most of the control to accessing this information. Downloadable music, instant messaging, and wireless voice and data are an integral part of the culture of the students who are the targets of the laboratory learning. If not the case already, it will become a student expectation. We need to focus it toward our teaching purposes and make it effective.

Comparing the effectiveness of traditional and discovery-based experiments. Morgan (1997) compared a discovery-based experimental procedure with a traditional one in which both used data collection devices. The 165 students were divided into two groups and took identical five-question quizzes and post-quizzes two months later. No differences were found between the quiz scores of the discovery versus traditional groups except possibly in the case of the question related to practical applications. Eighty-four percent of the discovery group answered it correctly, compared with 66% of the traditional students; the decrease in the percentage of those who answered correctly after 2 months was smaller for the discovery group (5.7%) than the traditional group (8.4%).

The availability of calculator-based labs (CBLs) and a microcomputer-based data collection system (MBL) in a secondary school setting gave Rigeman (2000) the opportunity to examine student learning under three different circumstances. Rigeman's studies examined the effect of the data collection methodology on students' understanding of the relationship between the pressure and volume of a fixed amount of gas at a constant temperature. Thus, 12 sections of students at a large urban high school were divided into three groups that generated the Boyle's Law data using a glass syringe and lead weights (i.e., the traditional group), using a CBL device, and using an MBL system. The latter two have similar pressure sensors but different display devices. A multiple regression analysis of student scores on the study instruments and their grade-equivalent scores from the Iowa Tests of Educational Development (ITED) Science, Quantitative Thinking, and Reading–Vocabulary subtests showed consistent correlation. However, no significant differences were found between the traditional and the CBL or MBL technology groups ($F = 0.44$, $p < 0.05$).

Meade (2002) examined several aspects of student learning in two general chemistry experiments conducted at a large midwestern university. Discovery or traditional versions of these experiments were prepared and provided to two groups of students who were divided based on several criteria used to differentiate their TAs. Quantitative measures included data from a conceptual pretest, a midterm exam, and an attitude survey with the same form being used as a pre- and post-survey. Qualitative data sources included a questionnaire administered after the treatment period and focus group questions. The goals were to assess the use of process skills, the effect of learning cycles on student attitudes, and reasoning skills. The results show that high-inquiry treatment students perceive that they used more process skills than their low-inquiry counterparts. For example, 60% of low-inquiry and 68% of high-inquiry students said that they could provide evidence, state conclusions, and make inferences. Only 29% of low-inquiry and 34% of high-inquiry students felt that they were encouraged to design their own experiments. A small focus group (4 low-inquiry and 9 high-inquiry students) did point to some differences in the ability of the laboratory to help them test ideas, to provide them with thinking skills that are useful outside of school, to answer questions, and to increase curiosity.

Designing a curriculum using EDC devices. So how does one make the transition from a set of traditional verification experiments to discovery-based ones using electronic data collection? In a large program, the place to start may be a pilot program. Both authors began by using three-year-old computers donated by campus information technology services. A dozen interface devices and sets of sensors get one into the business. Most vendors offer some level of on-site training in the use of the hardware and software, workshops at national meetings, or summer institutes on college campuses. Often an "expert" can be found at a neighboring institution. Such a support system may be necessary since many faculty members fear being embarrassed by not being well-versed in the use of the hardware or software. The authors have both used undergraduate and graduate students to develop new experiments and to accumulate institutional expertise.

Figure 5. **Students working with electronic data collection systems.**

Figure 5 shows students working at new stations at a western university. The workbench accommodates four students, in two sets of pairs. Group work is assigned to the pairs and the groups of four. The monitors are on the outside walls and therefore do not obstruct the view around the room. (EDC works equally well in 50-year-old labs that were not specifically designed for it.) Note that the benchtop holds the chemicals, the interfaces and sensors, and the student notebooks and manuals. Even with laptops among the wet chemistry, we have experienced little or no problems with spills or damage.

All of the commercial vendors offer sets of experiments for the high school or college level. Some will provide an entire curriculum. Some experiments have begun to appear in the chemical education literature. It is possible to start with a mix of traditional and discovery experiments but the students may become confused about the course goals. Likewise, the instructor may have difficulty separating which factors arise from different causes. Like many other aspects of teaching, becoming well-versed in the use of this methodology takes practice. Expect that the process will be formative—it will get better after several semesters and one should expect a transition and development period.

A "case" for a new curriculum. The University of Iowa has implemented a course redesign in general chemistry that attempts to integrate the "best practices" promoted in this section (Pienta et al., 2004). The laboratory portion of the course consists of two activities, case-study sessions and experimental sessions, that meet in alternate weeks. The material for the former involves a practical problem that serves as the motivation/introduction for the experiment that the students conduct the following week. An example is a calorimetry experiment involving the energy content of some food items; this experiment is illustrated by a case study outlining the practice regimen and caloric intake of a distance runner. The question asks whether the runner needs to eat a "PowerBar." The case-study sessions are limited to approximately 60–70 students, are conducted by an instructor (rather than a TA), and involve group discussions and activities concerning both the concepts and the procedural issues (including safety). The hands-on experimental session involves group work (generally data collection in pairs and pooling of data from the entire section of 20–24 students), electronic data collection and manipulation, a guided inquiry or discovery format (i.e., laboratory materials), and laboratory assessment (replacing the traditional laboratory report) that is clearly defined by rubrics. Training opportunities for teaching assistants about learning theory and procedural issues for a high-inquiry approach are being addressed but represent an ongoing struggle. Several successful models have been reported (Tien, Rickey, and Stacy, 1999; Varma-Nelson and Coppolla, Chapter 13).

Opportunities

For the future, laboratory instruction must continue to adopt effective pedagogical approaches (including group work, guided inquiry, and the science writing heuristic) together with the use of data collection systems. Alternative and appropriate assessments (including posters, laboratory practicals, and concept maps) must come with those changes in activity. In the words of Nakhleh, "It is no longer appropriate to simply assess laboratory work by reading laboratory reports or by testing individuals with exams that evaluate fact acquisition" (Nakhleh, Polles, and Malina, 2002, p. 85).

The chemical education research community must continue to study the practices that get implemented and seek unequivocal explanations. Some guidelines from Nakhleh, Polles, and Malina (2002) are given in a set of "implications for future research" that include

- the study of distributed cognition including data from case studies, field observation, and videotaping of laboratory sessions
- an analysis of the perceived properties of objects like instruments and devices for visualizing data [so-called "affordances" as defined by Pea (1993) and elucidated by Cole and Engestrom (1993)]
- the framework of learning and interaction with the environment
- student perspectives, including their views and goals and those of teaching assistants
- alternative assessments, particularly qualitative techniques.

The teaching laboratory is worth keeping and we, as instructors, can make it into an interesting and effective learning environment. At the same time, we as researchers, can provide the evidence of its successes and use the information to continue the process.

Suggested reading

Nakhleh, M. B., Polles, J., and Malina, E. (2002) "Learning chemistry in a laboratory environment" in *Chemical Education: Towards Research-based Practice*. J. K. Gilbert, O. De Jong, R. Justi, D. F. Treagust and J. H. Van Driel (Eds.). Dordrecht (The Netherlands): Kluwer Academic Publishers. pp 69–94.   This comprehensive review thoroughly covers the chemical education research literature related to the laboratory experience.

References

Abraham, M. R, Cracolice, M. R., Graves, A. P., Aldhamash, A. H., Kihega, J. G., Gil, J.G. P., and Varghese, V. (1997). The nature and state of general chemistry laboratory courses offered by colleges and universities in the United States. *J. Chem. Educ.* 74, 591–4.

Amend, J. R., Fursteneau, R. P., and Tucker, K. A. (1990). Student-designed experiments in general chemistry using laboratory interfacing. *J. Chem. Educ.* 67, 593–5.

Amend, J. R., Fursteneau, R. P., Howald, R. A., Ivey, B. E., and Tucker, K. (1990). An integrated workstation for the freshman laboratory. *J. Chem. Educ.* 67, 333–6.

Amend, J. R., Tucker, K. A., and Fursteneau, R. P. (1991). Computer-interfacing—A new look at acid-base titrations. *J. Chem. Educ.* 68, 857–60.

Amend, J. R., and Fursteneau, R. P. (1992). Employing computers in the nonscience-major chemistry laboratory. *Journal of College Science Teaching.* 110–4.

Amend, J. R., Morgan, M.E., and Whitla, A. (2000). Inexpensive Digital Monitoring of Signals from a Spectronic-20 Spectrophotometer.J. Chem. Educ. 77, 252–3.

Bailey, C. A., Kingsbury, K., Kulinowski, K., Paradis, J., and Schoonover, R. (2000). An integrated lecture-laboratory environment for general chemistry. *J. Chem Educ.* 77, 195–9.

Bates, G. R. (1978). The role of laboratory in secondary school science programs.In M. B. Rowe (Ed.) *What research says to the science teacher*. Washington, DC: National Science Teachers Association. 55–82.

Blosser, P. E. (1983). What research says: The role of the laboratory in science teaching. *School Science and Mathematics.* 83, 165–169.

Bondeson, S. R., Brummer, J. G., and Wright, S. M. (2001). The Data-Driven Classroom. *J. Chem. Educ.* 78, 56–7.

Brasell, H. (1987). The effects of real-time laboratory graphing on learning graphic representations of distance and velocity. *Journal of Research in Science Teaching.* 24, 385–95.

Brueningsen, C., and Bower, W. (1995). Using the graphing calculator in two dimensional motion plots. *Physics Teacher.* 33, 314–16.

Caniglia, J. (1997). The heat is on! Using the calculator-based laboratory to integrate math, science, and technology. *Learning and Leading with Technology.* 25, 22–7.

Choi, M. M. F., Wong, P. S., Yiu, T. P., and Case, M. Application of a Datalogger in Observing Photosynthesis. *J. Chem. Educ.* 79, 980–1.

Choi, M. M. F., Wong, P. S., Yiu, T. P., and Case, M. Application of a Datalogger in Biosensing: A Glucose Biosensor. *J. Chem. Educ.* 79, 982–4.

Cole, M., and Engestrom, Y. (1993). A cultural-historical approach to distributed cognition. In G. Salmon (Ed.) *Distributed cognitions: Psychological and educational considerations.* New York: Cambridge University Press. 47–87.

Cooper, M. M. (1994). Cooperative chemistry laboratories. *J. Chem. Educ.* 71, 307.

Cooper, M. M. (1995). Cooperative learning: An approach for large enrollment courses. *J. Chem. Educ.* 72, 162–4.

Cortes–Figueroa, J. E., and Moore, D.A. (2002). Using a graphing calculator to determine a first-order rate constant when the infinity reading is unknown. *J. Chem. Educ.* 79, 1462–4 (See also ibid., 76, 635–8).

Donato, H. (1999). Graphing calculator strategies for solving chemical equilibrium problems. *J. Chem. Educ.* 76, 632–4.

Durick, M. A. (2001). The study of chemistry by guided inquiry method using computer-based laboratories. *J. Chem. Educ.* 78, 574–5.

Fursteneau, R. P. (1990). Application of computers for experimental design, data acquisition, and analysis in the chemistry laboratory. Ph.D. dissertation, Montana State University.

Grimsrud, E., and Amend, J. R. (1979). Coulometry experiments using simple electronic devices. *J. Chem. Educ.* 56, 131–2.

Hickman, A. B., Helburn, R. S., and Delinger, W. G. (2000). Calculator-based instrumentation: The design of a digital temperature probe based on I2C technology. *J. Chem. Educ.* 77, 255–7.

Hodson, D. (1990). A critical look at practical work in school science. *School Science Review.* 70 (256), 33–40.

Kirschner, P.A., and Meester, M. A. M. (1988). The laboratory in higher science education: problems, premises, and objectives. *Higher Education.* 17, 99–119.

Lapp, D. A., and Cyrus, V. F. (2000). Using data-collection devices to enhance students' understanding. *Mathematics Teacher.* 93, 504–11.

Lazarowitz, R., and Tamir, P. (1994). Research on using laboratory instruction in science. In D. L. Gabel (Ed.) *The handbook of research on science teaching and learning.* New York: Macmillan. 94–128.

Meade, K. M. (2002). The effects of inquiry instruction on student learning in technology-based undergraduate chemistry laboratories. Ph.D. dissertation, University of Iowa.

Mokros, J. R., and Tinker, R. F. (1987). The impact of microcomputer-based labs on children's ability to interpret graphs. Journal of Research in Science Teaching. 24, 369–83.

Morgan, M. E. (1997). Application of computers for experimental design, data acquisition, and analysis in the chemistry laboratory. Ph.D. dissertation, Montana State University.

Nakhleh, M. B. (1994). A Review of Microcomputer-based labs: How have they affected science learning? *Journal of Computers in Mathematics and Science Teaching.* 13, 367–81.

Nakhleh, M. B., and Krajcik, J. S. (1994). The influence of level of information as presented by different technologies on students' understanding of acid, base, and pH concepts. *Journal of Research in Science Teaching.* 31, 1077–96.

Nakhleh, M. B., Polles, J., and Malina, E. (2002). Learning chemistry in a laboratory environmentIn *Chemical Education: Towards Research-based Practice*, J. K. Gilbert, O. De Jong, R. Justi, D. F. Treagust and J. H. Van Driel (Eds.). Dordrecht (The Netherlands): Kluwer Academic Publishers. pp 69–94.

Nemirovsky, R., Tierney, C., and Wright, T. (1998). Body motion and graphing. *Cognition and Learning*. 16, 119–72.

Nicol, M. (1997). How one physics teacher changed his algebraic thinking. *Mathematics Teacher*. 90, 86–9.

Pea, R.D. (1993). Practices of distributed intelligence and designs for education.In G. Salmon (Ed.) *Distributed cognitions: Psychological and educational considerations*. New York: Cambridge University Press. 47–87.

Pickering, M. (1993). The teaching laboratory through history. *Journal of Chemical Education*. 70, 699–700.

Pienta, N. J., Cannon C., Tardy, D., Larsen R., and El–Maazawi, M. (2004). Case-Study Sessions in the Laboratory Portion of a General Chemistry Course. In preparation.

Pienta, N. J., and Tardy, D. C. (2004). Private communication.

Reeves, J. H., and Ward, C. R. (2003). Private communication.

Ricci, R. W., Ditzler, M. A., Jarret, R., McMaster, P., and Herrick, R. (1994). The Holy Cross Discovery Chemistry Program. *J. Chem. Educ.* 71, 404–5.

Rickey, D., and Stacey, A.M. (2000). The Role of Metacognition in Learning Chemistry. *J. Chem. Educ.* 77, 915–20.

Rigeman, S. A. (2000). The impact of technology on chemistry students' construction of meaning from a laboratory investigation of Boyle's Law. Ph.D. dissertation, University of Iowa.

Roser, C. E., and McCluskey, C. (1999). Lightstick kinetics. *J. Chem. Educ.* 76, 1514–5.

Roth, W. M. (1994). Experimenting in a constructivist high school physics laboratory. *Journal of Research in Science Teaching*. 31, 197–223.

Rudd, J. A., Greenbowe, T. J., and Hand, B. M. (2001).Using the Science Writing Heuristic to Move toward an Inquiry-Based Laboratory Curriculum. *Journal of Chemical Education*. 78, 1680–6.

Rudd, J. A., Greenbowe, T. J., and Hand, B. M. (2002). Recrafting the General Chemistry Laboratory Report. *Journal of College Science Teaching*. 31, 230–4.

Sales, C. L., Ragan, N. M., and Murphy, M. K. (1999). Creative uses for calculator-based (CBL) technology in chemistry. *NEACT Journal*. 18, 16–9.

Tien, L.T., Rickey, D., and Stacey, A.M. (1999). The M.O.R.E. Thinking Frame: Guiding Students' Thinking in the Laboratory. J. College Science Teaching. 318–24.

Toothacker, W. S. (1983). A critical look at introductory laboratory instruction. *American Journal of Physics*. 51, 515–20.

Vendors (2003). Production of interfaces and sensors by a popular vendor began in 1981 with a physics teacher in his garage: *www.vernier.com* (Vernier). Additional commercial ones have appeared: *www.pasco.com* (Pasco);*www.microlabinfo.com* (Microlab); *www.measurenet-tech.com* (Measurenet); *www.scitechnologies.com* (LabWorks). Labview is a very elaborate and powerful software package that enables control and design of experiments but is more likely to be the system of choice for scientific research because of the learning curve and expense.

Viswanathan, R., Lisensky, G., and Dobson, D.A. (1996). Off-the-shelf portable data acquisition: Interfacing a serial-port-equipped multimeter. *J. Chem. Educ.* 73, A41.

Vitz, E., and Egolf, B. P. (2002). LIMSport: Optimizing a Windows-based computer data acquisition and reduction system for the general chemistry laboratory. J. Chem. Educ. 79, 1060–2 and references cited therein.

15

Wireless Inside and Outside the Classroom

Jimmy Reeves and Charles R. Ward
Department of Chemistry
University of North Carolina at Wilmington

Abstract

For the past two decades, the chemical education community has explored the use of hypermedia technology and networked computing in teaching, with many pundits predicting dramatic changes in our methods of instruction as the "anywhere, anytime" learning revolution takes place. The promises of this revolution are indeed enticing. Interactive learning modules, animations, and videos would help bring chemistry to life, giving students glimpses of the nanoworld of atoms and molecules that scientists carry around in their heads. Virtual classrooms would provide anywhere, anytime access, facile communication and rich data sharing. Students would be linked to one another and to their instructors, providing unique new opportunities to engage in cooperative learning activities, as well as new avenues to assess student learning.

Biographies

Jimmy Reeves is an associate professor of chemistry at the University of North Carolina at Wilmington and has researched the application of technology to chemistry instruction for the past thirteen years. He was the principle author of *ActivChemistry*, an interactive chemistry simulation published by Benjamin Cummings, and coauthor of *Kitchen Chemistry*, a set of general chemistry labs for distance-learning students designed to be performed at home, soon to be published by Prentice Hall. He served four years on the ACS General Chemistry examination committee, and organized and presented over 100 seminars at regional and national ACS meetings.

Charles Ward is Professor and Chair of chemistry at the University of North Carolina at Wilmington. He has been actively studying the use of technology in teaching chemistry since 1978. He was the co-author, along with Jimmy Reeves, of *Microcomputer Applications in Chemistry*, a textbook for teaching the fundamentals of computer use in chemistry originally published by Scott-Foresman. During his tenure as head of the information technology program at UNCW, he developed an interest in mobile computing devices and their application to science and mathematics instruction.

Although a wealth of research suggests that actively involving students in their learning improves their understanding, many instructors hesitate to adopt approaches such as small-group learning or guided-inquiry because they consider the logistics required for their implementation to be prohibitive. Others believe that students are not capable of learning the subject matter if all of the material is not presented in lecture. Furthermore, their teaching evaluations, which often include a question that asks whether they "present clear, well-organized lectures," can suffer from student expectations that they are "paid to teach them" as opposed to helping them learn. Virtual applications that facilitate interactive group work and anonymous communication with the instructor provide a potential solution, but they require that all students have devices that can access the applications. These devices have historically been desktop or laptop computers that require large amounts of desk space, have long boot-up times, and require electrical power and wired network/internet connectivity, making them impractical for most large classrooms. Moreover, even when classrooms are "wired" and students are required to purchase identical laptop computers, many fail to carry them to class. Finally, tools that facilitate virtual interactions are often cumbersome and difficult to use.

Despite these problems, the revolution may not be far off. Advances in wireless networking and hand-held computing that have occurred over the last few years are providing the tools to overcome many of the impediments to progress. These technologies are paving the way for the creation of a learning environment that is truly anytime and anywhere, as available in the lecture hall as it is at home or in specialized computer labs. In this chapter we describe the potential educational implications of a *Mobile Learning Environment*, along with preliminary data that demonstrate the effectiveness of the approach and outline the technological challenges that must be addressed to enable its creation.

Background

It is anticipated that, in the near future, students at campuses across the United States will make extensive use of mobile computing devices that they will carry with them at all times. Mobile devices such as Pocket PCs, Tablet PCs, laptops, and cell phones are more popular than ever (Asay, 2002). A number of U. S. campuses, including the University of Minnesota–Crookston, Wake Forest University, Stanford University, Des Moines Area Community College, and California State University–Monterey Bay, are requiring such devices while hundreds are testing the feasibility as pilot programs (Boettcher, 2001; Dominick, 2002). More than 77% of U. S. colleges and universities report some use of wireless networks while approximately 14% report having wireless networks providing seamless network access within the physical boundaries of their campuses (Fallon. 2002). The University of Georgia is one of several institutions that have extended their campus wireless access to include large areas of the surrounding community (University of Georgia, 2002).

Of the colleges and universities investigating the use of the mobile devices, some are simply aiming to "get the technology into students' hands" and plan to find instructional uses at later dates (University of Louisville, 2003). Others are finding that the lack of software limits the devices to simple operations that do not capitalize on the benefits of the mobile form factor (Fallon, 2002). In our research, we have found that the limited output options for mobile devices prevent students from easily printing assignments or submitting them electronically, since both involve multi-step processes that require the use of additional applications as well as an understanding of how these applications work together. Electronic submissions also create issues for instructors, including how to store, access, grade, and return files to the students. Moreover, this form of electronic submission is not easily scalable to multiple instructors and class sections. Other limitations of mobile computing devices include minimal input and output (I/O) options, small screen size, and potential loss of files with loss of battery power. While these problems are not insurmountable, they do require time, effort, and creative thinking to resolve.

Attempts to use mobile devices in campus settings have met with mixed success. For example, the University of Louisville's Health Sciences Center requires all incoming medical and dental students to own a PDA. However, their solution for providing students with content and new applications is through HotSync stations distributed around campus (University of Louisville, 2003). If students are unable to access a hotsync station because the unit is down for repair, other students are waiting in line, or it is too far out of the way, then they will not be able to transfer files or access the latest information about their courses. By contrast, an MLE should provide direct *drag-and-drop* methods for downloading content and applications over the wireless network and time-critical information should be *pushed* to students' devices with no need for special software. Thus, it is critical that the overarching problems facing mobile computing now and in the future be addressed by creating an environment that is flexible, easy to use, and conducive to campus-wide collaborative communication anytime, anywhere. To design this environment, we are investigating the educational advantages and limitations of a variety of wireless devices, and how each device might impact current educational initiatives designed to promote active learning and collaboration among our students.

Affecting the Lecture

Much of the debate surrounding the use of lecture-based science classes has focused on the passive role assumed by students in lecture, where communication has traditionally flowed from instructor to student (Cooper, 1995; Kraft, 1985; Ebert–May, 1997). Numerous studies have shown that student performance in science classes improves with increasing levels of active participation by students in classroom discussions (Francisco, 1998; White, 1972; Hunter, 1973). Cooperative learning techniques such as Process Oriented Guided Inquiry Learning (POGIL), peer instruction using "ConcepTests" and "turn-to-your-neighbor" activities

(Cooper, 2000; Fagen, 2002; Borman, 2003) are designed to encourage communication from students to instructors and among students themselves. They devote anywhere from part of the lecture to the entire lecture to guided discussions among small groups of students and often require that the instructors modify their role as "sage on the stage" to "guide on the side." Potential benefits that wireless mobile devices can bring to these cooperative techniques include streamlined communication between the groups and the instructors within the classroom and facile virtual group interactions outside the classroom.

Despite the documented successes of cooperative learning techniques, many instructors who teach large introductory lectures are reluctant to implement small-group activities and attempt instead to involve the students by holding large-scale classroom discussions. Often the instructor will ask a multiple-choice or yes–no question and solicit a "show of hands" of students in response to each choice. The problem with this approach is that participation is often limited to a small number of students willing to risk being wrong in front of the entire class. A modification that involves holding up colored cards to indicate student choices (red for A, blue for B, etc.) used by Tom Greenbowe at Iowa State University (private communication) has succeeded in improving participation, possibly because the instructor is the only one who can easily see the students' responses, preserving their anonymity. He found that the responses provided a snapshot of how the class as a whole understood the concept underlying the question, giving him instant feedback with which to decide how to proceed with their instruction. One problem with this approach is that the information is lost as soon as the students lower the cards.

Recently, a technological approach to collecting student responses, which originated with 1950s TV game shows (Audience Response Systems), has gained renewed popularity. Products such as the *Classroom Performance System* (*EInstruction,* 2004) and *Turningpoint* (*TurningPoint,* 2004) collect and display student answers provided by wireless input devices (including wireless PDAs and TabletPCs) that the students bring to class. After the instructor poses a preloaded question, students broadcast their responses to a database, where they can be accessed by the instructor in a number of different formats. Many of the devices limit the responses to multiple-choice and yes–no questions, though alphanumeric units are also available. The instructor can instantly display the number of responses for each choice and the identity of the students who have responded. Many of the remote control devices use infrared radiation, so they can only transmit data if pointed directly at a detector. This can be a problem, especially in very large rooms. Also, the number and location of detectors needed to provide access to every student should be carefully considered before systems using infrared technology are implemented.

Using Wi-Fi-enabled PDAs or TabletPCs to provide student responses has the potential to expand both the kinds of questions that can be asked and the types of answers that can be transmitted. A prototype of such a system has been in use at UNCW for two years. Designed to be part of the Mobile Learning Environment, the Numina II Student Response System incorporates the following features.

- *Web-based.* The system relies on Web technologies for all communications that take place between the system and users. This means that any device capable of displaying a simple webpage can be used to interact with the system.
- *Hardware independent.* The system is independent of the hardware platform available to students and instructors and will work with most computing technologies that schools already own, including PCs, Macs, workstations, laptops, desktops, Pocket PCs, Palm devices, and Web-enabled mobile phones.
- *OS independent.* The system is independent of the operating system present on devices so it will work with Windows, Mac OS, Palm OS, Pocket PC OS, UNIX, and Linux. (Certain advanced features require the operating system to support Macromedia Flash.)
- *Utilizes existing network infrastructure.* The system operates over wired as well as wireless networks using whatever network infrastructure already exists on campus.
- *Multiple response interfaces.* The availability of a rich variety of interfaces means that students are not limited to simple multiple-choice or yes–no responses.
- *Database driven.* All data related to a session, regardless of whether questions were presented from the database or "on the fly," are stored in an online database for immediate display and analysis and for review at a later time.

- *Student anonymity.* Classroom management data (e.g., attendance) are stored separately from student response data to ensure the anonymity of student responses.

When using the Numina II SRS system, the instructor presents a question or series of questions to the class. The questions can be drawn from a database or asked on-the-fly. Each question is displayed to the entire class using a large monitor or projection system as shown in Figure 1.

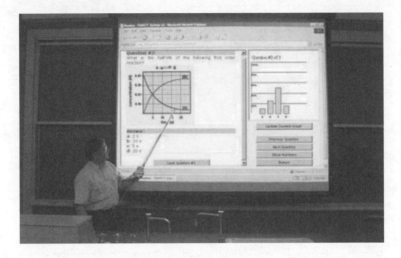

Figure 1. SRS instructor and classroom view.

The format of the student response depends on the interface *chosen by the instructor* for the particular question being asked (i.e., multiple-choice, yes–no, true–false, confidence, graphics, etc.). This interface is displayed on the student's device, which may be a desktop, laptop, Pocket PC, or cell phone. Examples of student response interfaces are shown in Figure 2.

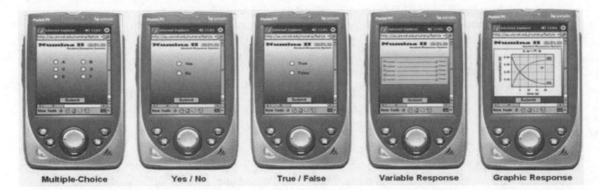

Figure 2. SRS interfaces shown on Pocket PCs.

The response submitted by the student is stored in a database and then is immediately summarized in graphical format. The instructor has the option of showing the summary as student responses are being received or waiting until all of the students have finished responding. The most common technique is to wait until everyone has finished, avoiding the problem of slow responders being influenced by the results of those who responded quickly.

The format for the display of student data depends on the type of interface that was chosen. For example, data from the multiple choice response interface are displayed as a bar chart (Figure 3), whereas data from the graphics response interface are displayed as a scatter chart.

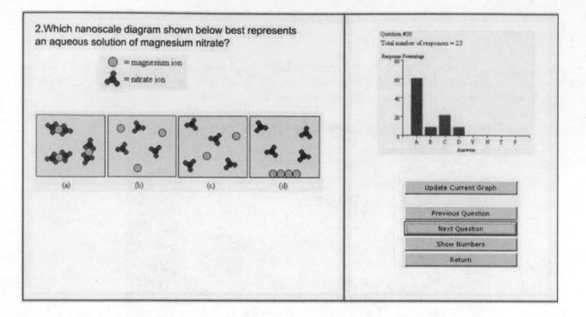

Figure 3. Question and responses with multiple-choice interface.

The use of a Student Response System in lecture and laboratory classes has resulted in noticeable improvements in a number of important areas:

- Student participation in question sessions is consistently near 100%.
- Students who might normally monopolize the discussion of a question must wait until all students have had a chance to provide a response.
- Student–instructor interaction, in the form of discussion, is increased (frequency) and more widespread (distribution) when a question is presented with the SRS system.
- Instructors gain immediate information regarding the extent to which students understand the concepts or procedures being presented.
- Instructors make informed decisions based on student data that impact the pace of class, the curriculum, and classroom procedures.
- Nearly 100% of participating students report that they prefer the SRS system to normal classroom questioning techniques.

Mobile Devices in the Classroom

Although the Numina II SRS system was designed to work with many different types of computing equipment, much of the design work focused on making the system compatible with mobile devices such as Personal Digital Assistants (PDAs) and Web-enabled mobile phones. Many instructors believe that using PDAs and mobile phones in class will lead to disruptive behavior that might compromise the instructional value of the SRS system. The following three questions are the ones most frequently asked when discussing classroom use of a Web-enabled SRS system:

1. Does the distribution and collection of the devices take a significant amount of class time?
2. Do technical problems with the devices and wireless network connections interfere with the implementation of the system?
3. Does off-task behavior increase when students have mobile Internet-ready devices in their hands?

To address the first of these questions, time management studies were conducted in a large lecture hall and in several smaller laboratory sections (Ward 2003). It was found that it took less than four minutes to distribute and collect 100 PDAs in the lecture hall and less than three minutes to distribute and collect 24 PDAs in the lab.

Furthermore, most of the distribution and collection occurred during class changes so that, on average, less than two minutes of instructional time was lost.

To address questions two and three, trained evaluators observed two laboratory sections of general chemistry while the lab instructors conducted post-lab sessions using the SRS system. Only two technical problems were found during the sessions. Neither affected the entire class and both were resolved in less than two minutes. Moreover, in both sections, they found that over 80% of the students were on task during the entire session and over 90% for at least half of the time.   The most frequently observed off-task behaviors were Web surfing, reading e-mail, and playing solitaire. These results matched anecdotal information from other instructors and confirmed that technical issues and off-task behavior related to PDAs and wireless networking have a negligible impact on instruction when the students are engaged in active learning tasks.

Other Features of Wireless Technologies

In addition to facilitating anonymous student feedback, wireless devices can allow students to

- *Take tests online.*
  To assess the effects of new educational technologies on student learning, tests must include questions that probe the concepts taught using the technology. If the instructor uses a video to illustrate a concept, a wireless device can make the video available to the students as they are answering questions about the concept. This represents a significant advantage over traditional exams which must rely on static, black-and-white diagrams and graphs. This places severe limitations on the kinds of questions that can be asked. .

- *Download up-to-date course notes.*
  Many instructors provide students with their course notes, often in the form of handouts (course packs) or computer files. Using wireless devices, instructor notes can be downloaded immediately before class and used as an outline for taking notes during class.

- *Acquire and share experimental data.*
  Wireless devices can acquire experimental data and send it to a database where it can be shared with instructors and other group members. Group members can work in virtual collaborations to analyze data and write lab reports.

- *Hold virtual meetings.*
  Using wireless devices, students can engage in audio/video conferences with one another (many to many) and with their instructors (many to one), on or off campus.

- *Locate nearby resources and people.*
  "Location-aware services" pinpoint the whereabouts of users logged into the network and networked resources such as printers.

Tools for the Wireless World

Although many who follow advances in educational technology expected that a single device would emerge to serve all of the students' technology needs, experience has demonstrated that a one-size-fits-all policy for devices is ill advised, at least in the short term. Early adopters of mandatory laptop ownership, which put identical laptops in the hands of all students and faculty, found that many students failed to bring their computers to class, and those that did had to cope with long boot-up times and short battery lives. On most of today's campuses, it is much more common to see students carrying cell phones than laptop computers.

One device that students *are* likely to carry to class is a Tablet PC. Tablet PCs are wireless laptop computers designed for note taking. Using an attached stylus, students can write directly on the tablet, and sophisticated character recognition software is available to translate their words into standard text. Because Tablet PCs are lighter than laptop computers, with longer battery lives and shorter boot-up times, it's much more likely that students will find them to be desirable replacements for laptops. However, students are not likely to carry Tablet PCs when they are not on their way to class. Instead, they will have a slightly less powerful computer with a much smaller footprint, such as a Pocket PC or PDA. Capable of running much of the same software (including sophisticated molecular-modeling programs) as a desktop computer, these devices represent an important bridge between cell phones (which have very small screens and minimal computing capabilities) and full-featured computers (like laptops and Tablet PCs). The PDA screen is a convenient size for viewing email messages and websites, especially when web pages are formatted to accommodate them. Expansion slots can accept additional memory, Global Positioning System (GPS) devices, data acquisition units, cameras, and a host of other add-ons. Software (usually under $40) turns them into advanced scientific calculators, video and audio players, remote control devices, and cell phones, though their size may be too big to be a practical replacement for the small, sleek phones popular today. Because the operating system is contained in ROM, the devices are "instant-on" and have long battery lives.

Each type of device has benefits and limitations, and it is likely that many students will own all three (Tablet PC, PDA, and a cell phone) in the near future. To fully exploit their educational potential, it's critical that we determine the educational applications and activities that are appropriate for each device and that we create an operating environment that provides a consistent, intuitive interface for all devices, as well as common functions for carrying out key operations such as file sharing and communication.

The Current Operating Environment

The diagram in Figure 4 depicts a hierarchical view of existing computing services as well as those we would like to see developed. Individual software services such as FTP and e-mail are ranked under the *Disaggregated Services* category. These represent stand-alone applications and services with proprietary interfaces, limited interoperability, and awkward or non-existent data interchange capabilities.

Multiple service protocols, typically requiring a common login (such as portals), are included in the *Aggregated Services* level. These applications and services, such as *WebCT* and *BlackBoard*, overcome some of the shortcomings of the items in the *Disaggregated Services* level, but they still fail to be workable solutions for the problems facing mobile computing. They do not, for example, provide rich data sharing and location aware services, and multi-way communication is limited to text. Moreover, information stored by these products is often difficult to integrate with the data on the instructor's computer. Finally, they are not designed for mobile devices, the most promising technology for anytime, anywhere learning. What is required in a mobile computing environment is a layer of *Integrated Services* for which no solution currently exists.

	INTEGRATED SERVICES	• Multi-way communications • Teamwork • Rich data sharing • Personalized app-level access controls • In-depth/broader access • Expanded physical access
	AGGREGATED SERVICES	• Portals (Campus Pipeline) • E-learning Systems (WebCT, Blackboard, Lotus Learning Space) • Real-time Collaboration Applications (NetMeeting, nEwhere, FirstClass) • myStanford
	DISAGGREGATED SERVICES	• Individual protocols, ports, or applications • Examples: Forums, FTP, Email, Web, Chat/IM, SIS, Telnet, SRS, Excel, Word

Figure 4. Levels of Aggregation of Technology Services.

A Model for a Mobile Learning Environment

An interface similar to that shown in Figure 5 might be envisioned for any of the devices used in a Mobile Learning Environment (MLE). The options available would depend on the type of device, but the operations required to exercise any option would be consistent from device to device.

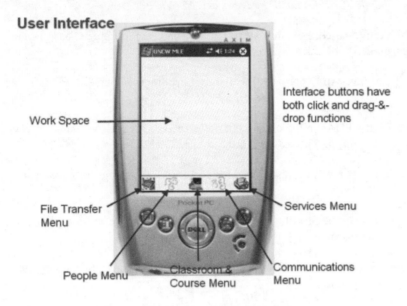

Figure 5. Mobile Learning Environment Interface.

When students log into the MLE, information such as their class schedule, instructors, and study groups would automatically be accessed from a database. They would be immediately alerted to newly posted class information and communications from classmates or instructors. If, for example, a student wanted to share laboratory data with the other members of her group, she would tap on the *File Transfer Menu* icon, producing a list of files available for sharing from the student's directory. Once the appropriate files are chosen, a list of potential recipients would appear. Choosing recipients from this list provides them with access to the files. A similar operation would permit the student to print a document on a network printer chosen from a "location aware" campus map, participate in a videoconference help session with one of her instructors, or answer questions posed during lecture using the SRS system. Such a design would facilitate student participation in learning both inside and outside the classroom by providing an interface specifically designed to handle their educational needs while removing technical barriers that intimidate and discourage their use of technology. Such an environment would be an important bridge to the learning anytime, anywhere vision that has tantalized the science education community for years.

Suggested Reading

Bates, A. W., and Poole, G. (2003). *Effective Teaching with Technology in Higher Education: Foundations for Success.* San Francisco: Jossey-Bass.

Curtis, M., Gramling, A., Reese, K., Wieczorek, A., Norris, C., and Soloway, E. (2004). *Pocket PC Computers—A Complete Resource for Classroom Teachers.* Washington, DC: ISTE.

Alesso, H. P., and Smith, C. F. (2001). *The Intelligent Wireless Web.* Boston: Addison-Wesley.

References

Asay, P. (April 2002). Tablet PCs: The killer app for higher education. *Syllabus.* 15, 40–41 Available online at *http://www.syllabus.com/article.asp?id=6246*. Last accessed February 11, 2003.

Boettcher, J.V. (June 2001). The spirit of invention: Edging our way to 21st century teaching. *Syllabus*. 14, 10–13. Available online at *http://www.syllabus.com/article.asp?id=3687*. Last accessed February 11, 2003.

Borman, Stu (March 2003). Nontraditional Teaching: Online Conference Focuses on Methods that Enhance or Replace Conventional Lecture. *Chemical and Engineering News*. 81, 45–47.

Cooper, J. L., and Robinson, P. (2000). Getting Started: Informal Small-Group Strategies in Large Classes. *New Directions in Teaching and Learning*. 81, 17–24.

Cooper, M. M. (1995). Cooperative Learning: An Approach for Large Enrollment Courses. *Journal of Chemical Education*. 72, 162–164.

Dominick, J. (September 2002). Ready or not—PDAs in the classroom. *Syllabus*. 16, 30–32. Available online at *http://www.syllabus.com/article.asp?id=6705*. Last accessed May 13, 2004

Ebert–M. D., Brewer, C. (1997). Innovation in Large Lectures—Teaching for Active Learning. *Bioscience*. 47, 601–608.

EInstruction. (2004). Classroom Reporting System, EInstruction Corporation. Available online at *http://www.einstruction.com/*. Last accessed on May 13, 2004.

Fagen, A. P., Crouch, C. H., and Mazur, E. (2002). Peer Instruction: Results from a Range of Classrooms. *Physics Teaching*. 40, 206–209.

Fallon, M.A.C. (November 2002). Handheld devices: Toward a more mobile campus. *Syllabus*. 16, 10–15 Available online at *http://www.syllabus.com/article.asp?id=6896*. , Last accessed May 13, 2004.

Francisco, J. S., Nicoll, G., and Trautmann, M. (1998). Integrating Multiple Teaching Methods into a General Chemistry Classroom. *Journal of Chemical Education*. 75, 210–213.

Hunter, W. E. (1973). Individualized Approaches to Chemistry versus Group Lecture Discussions. *Journal of College Science Teaching*. 2, 35–38.

Kraft, R.G. (1985). Group-Inquiry Turns Passive Students Active. *College Teaching*. 33, 149–154.
Tom Greenbowe, private communication.

TurningPoint. (2004). Thomson Brooks–Cole. Available online at *http://www.wadsworthmedia.com/TurningPoint/TurningPoint_Demo.html*. , Last accessed May 13, 2004.

University of Georgia: Wireless cloud permeates Athens. (November 2002). *Syllabus*. 16, 28–30. Available online at *http://www.syllabus.com/article.asp?id=6909*. Last accessed May 13, 2004.

University of Louisville: Med schools integrate handhelds. (February 2003). *Syllabus*. 16, 35–36. Available online at *http://www.syllabus.com/article.asp?id=7261*. Last accessed May 13, 2004.

Ward, C. R., Reeves, J. H., and Heath, B. (April 2003). Encouraging Active Student Participation in Chemistry Classes with a Web-based, Instant Feedback, Student Response System, NON-TRADITIONAL TEACHING METHODS: Methods Other Than Lecture and Assessment of these Methods. *Online Conference for ConfChem*.Available online at *http://www.chem.vt.edu/confchem/2003/a/ward/ConfChem_SRS.htm*. Last accessed April 22, 2004.

White, J. M. (1972). Freshman Chemistry Without Lectures: A Modified Self-Paced Approach. *Journal of Chemical Education*. 49, 772–774.

16

Using Multimedia to Visualize the Molecular World: Educational Theory Into Practice

Roy Tasker
School of Science, Food, and Horticulture
University of Western Sydney

Abstract

We know that many misconceptions in chemistry stem from an inability to visualize structures and processes at the molecular level. We also know that you cannot change a student's mental model of this level by simply *showing* him or her a different, albeit better, one. Using an evidence-based, multimedia information-processing model, the constructivist *VisChem Learning Design* probes a student's mental model of a substance or reaction before showing 3-D molecular animations portraying the phenomenon. In this way, students are focused on identifying the "key features." Molecular-level animations are compelling and effective learning resources, so they must be designed and presented with great care to avoid generating or reinforcing misconceptions.

Biography

Roy Tasker discovered the joy of chemistry as an undergraduate at the University of Queensland; in 1978, he moved to the University of Otago to complete his Ph.D. on Co(III)-mediated peptide synthesis. Following a number of short appointments, he moved to the University of Western Sydney in 1985, where he is currently an associate professor of chemistry. His major teaching contribution is at the freshman level, but he also teaches at advanced levels in inorganic chemistry and chemical education. Career highlights include the Royal Australian Chemical Institute (RACI) Chemical Education Division Medal (2002), one of only six awarded in its 20-year history, and the inaugural UWS Nepean Award for Excellence in Tertiary Teaching (1992).

His current R and D interests are in e-learning using interactive multimedia. Since 1996, he has worked in partnership with a multimedia production company, *CADRE design* (*http://cadre.com.au*), and conducted research with students on the effectiveness of the outcomes. Up to 2002, this partnership has produced nine multimedia programs on CD and the Web for WH Freeman and Co. to supplement and complement their chemistry and biochemistry textbooks. Emphasis has been on producing learning objects and designs to assist students in visualizing the molecular level and then to link their mental visual models to observations at the laboratory level and to the abstract symbolic/mathematical level. Many of the molecular-level animations used in these resources were produced in his *VisChem* project (*http://vischem.cadre.com.au*).

Introduction

Chemistry involves interpreting *observable* changes in matter at the concrete **laboratory level** (e.g., color changes, smells, and bubbles) in terms of *imperceptible* changes in structure and processes at the imaginary **molecular level** (atoms, molecules, and ions). These changes are then represented at an abstract **symbolic level** in two ways: qualitatively (using specialized notation, terminology, and symbolism), and quantitatively (using mathematics, such as equations and graphs). The need to be able to move seamlessly between these three "thinking levels," first described by Johnstone (1982, 1991), is a major challenge for students learning chemistry (Kozma and Russell, 1997). This subject is also discussed in Chapter 7.

Figure 1 is an example of how a chemistry instructor might provide students with the three levels of representation for an iron(III) thiocyanate equilibrium. The unchanging solution in the beaker at the laboratory level is shown with an animation portraying the dynamic, but imperceptible, processes at the molecular level. The equation represents the equilibrium at the symbolic level. Showing an animation of this equilibrium at the molecular level should help students build a better conceptual understanding of what it means for a system to be at equilibrium.

Figure 1. Chemical equilibrium presented at the three thinking levels.

Figure 2. Using the three thinking-level approach in the lecture theatre. This approach is also reinforced in the laboratory notes, tutorials and assessment.

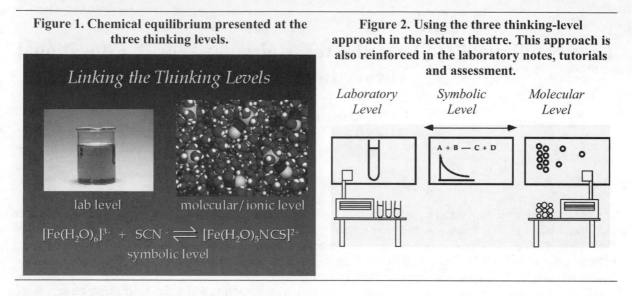

Various researchers have recommended teaching about the different levels of thinking in an explicit way, and helping students to draw links between the levels (Tasker, 1992; Tasker, Chia, Bucat, and Sleet, 1996; Russell, Kozma, Jones, Wykoff, Marx, and Davis, 1997; Hinton and Nakhleh, 1999). I first used these levels *explicitly* in my chemistry teaching in the late 1980s (Tasker, 1992), allocating different parts of the lecture stage to different levels (Figure 2), rewriting laboratory manuals, and designing exam questions to encourage students to integrate laboratory work and theory at each level.

However, I was frustrated by my inability to help students construct useful mental models of the molecular world. Due to a shortage of high-quality, compelling resources that portrayed the molecular level, most of my teaching was restricted to the laboratory and symbolic levels, in the hope that the students' models of the molecular world would "develop naturally". Students were left to construct these models from the static, often oversimplified, two-dimensional diagrams in textbooks; or from confusing ball-and-stick models; or from their own imaginative interpretation of chemical notation—for example, *Did "NaCl(aq)" mean that ionic solutions contained dissolved "NaCl molecules"?*

Since the mid-1970s, there has been a growing research literature with convincing evidence (de Vos et al, 1990, and references therein; Kleinman, 1987) that many student difficulties and misconceptions in chemistry resulted from inadequate or inaccurate models at the molecular level. Moreover, many of the misconceptions were common to students all over the world, and at different educational levels, and even among students who were performing well in formal examinations (Nurrenbem and Pickering, 1987; Nakhleh, 1993, 1992; Nakhleh and Mitchell, 1993; Niaz, 1995). The most important finding was that many misconceptions were extraordinarily resistant to change.

The purposes of this chapter are to

- present a multimedia information-processing model to guide your learning designs for helping students construct useful mental models at the molecular level.
- illustrate, with an example, one such design—the *VisChem Learning Design*—for developing a molecular-level model of aqueous ionic solutions.
- suggest how you can probe a deep understanding of concepts at the molecular level

- describe some of the messages, and misunderstandings, that can be communicated with animations

A model for learning through molecular-level visualization

The molecular world is multiparticulate, dynamic, and crowded, and the interactions are subtle and often complicated. Interactive multimedia offers the greatest potential for helping students to construct and apply useful mental models of this world. However, effective use of this technology for meaningful learning should be based on a theory that is evidence-based and able to inform teaching practice.

A cognitive theory of multimedia learning has been developed by Mayer (1997), and it is used to decide whether or not, when, and for whom, multimedia instruction is effective. Recently, it has been used to derive instructional design principles for multimedia explanations (Mayer and Moreno, 2000). This theory has much in common with the information-processing model developed by Johnstone (1997) and applied extensively to inform educational practice. A combination of aspects of these theories can form a useful basis to guide our use of multimedia to improve our students' visualization skills. An attempt to combine these theories is depicted in Figure 3.

Figure 3. A multimedia information-processing model for learning from audiovisual information. This is a composite of theoretical models proposed by Mayer (1997) and Johnstone (1997).

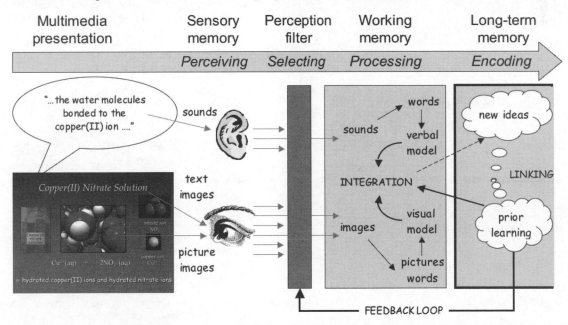

The theoretical stages in learning from audiovisual multimedia are summarized briefly below, together with their implications for teaching practice:

Stages in Construction Process	Implications for Teaching with Multimedia
Receiving multimedia information • perceiving verbal (sounds of words) and visual information (text and pictures) using the sensory receptors aroused by the attention networks in the brain	• activates the attention networks in the brain with an engaging context or problem • takes care with confusing language (e.g., "clear" versus. "colorless") and explains 3-D perspective structure conventions

Selecting information by the perception filter	
• selecting relevant information from the sensory receptors *on the basis of prior experiences, pattern recognition, and prior learning in the long-term memory*	• recalls prerequisite learning to prepare the perception filter by asking questions and encouraging students to communicate their preconceptions
Processing in working memory	
• storing and processing the selected information in short-term memory with *limited capacity* (7±2 "bits" of information)	• reduces extraneous, audiovisual clutter to maximize the "signal-to-noise" ratio and avoids memory overload
	• presents information in "chunks" (e.g., screen grabs) before combining them to experience the global perspective
	• enables the students to have *manual* control over animations (by dragging the PLAY bar) to reduce information presented per unit time, and to encourage cause–effect reflection
	• keeps the semantic content of the verbal and visual information the same, not just related, and certainly not conflicting
	• encourages group learning to exploit multiple working-memory capacities
• organizing the selected information into visual and verbal models *simultaneously*	• presents text and illustrations contiguously on the same page, and presents animations with simultaneous narration
	• presents concurrent narration rather than concurrent text
• integrating verbal and visual models into a coherent whole with input from prior knowledge	• uses familiar representations, and symbol keys
Encoding in long term memory	
• storing new ideas as separate or interconnected fragments in the long-term memory	• encourages meaningful learning by linking new ideas to as many existing ideas as possible, maximizing interconnections
	• provides opportunities for new ideas to be retrieved and applied in new situations
	• enables students to articulate and demonstrate to themselves and their peers what they are learning

Theory into practice: the *VisChem Learning Design*

This theoretical model, with its implications for good practice, informed the design of the constructivist *VisChem Learning Design* (Tasker, 2002) to develop the skill of molecular-level visualization of aqueous ionic solutions. This is one of a collection of exemplary ICT-based learning designs selected to facilitate the uptake of innovative teaching and learning approaches in Australian universities (Harper, 2001).

The design is described below and is illustrated with an example in the hope that you will test it, and/or develop your own designs to promote visualization of the molecular level. More details can be found on the website for the design (Tasker, 2002), together with a freely-downloadable construction tool and selected *VisChem* animations.

The *VisChem Learning Design* can be used in any chemistry topic that requires an accurate mental model of the molecular world. A typical learning experience in a face-to-face lecture context would involve students

- **observing** a chemical phenomenon (chemical reaction or property of a substance) as a lecture demonstration, lab activity, or audiovisual presentation; and **documenting** their observations in words and/or diagrams
- **describing** in words and **drawing** a representation of what is occurring at the molecular level to account for the observations, with the aid of the lecturer explaining the need for drawing conventions (e.g., relative size, movement, number, and crowding of molecules)
- **discussing** their representation with a peer, with the aid of the lecturer's advice to focus on the *key features* of the representation that explain the observations
- **viewing** an animation portraying the phenomenon at the molecular level, first without, then with, narration by the lecturer, and **looking** for key features that might explain the observations
- **reflecting** with the peer on any similarities and discrepancies between their own representations and the animation, and then **discussing** these with the lecturer
- **relating** the molecular-level perspective to the symbolic (e.g., equations, formulas) and mathematical language used to represent the phenomenon
- **adapting** their mental model to explain a similar phenomenon with an analogous substance or reaction

The key criteria for the success of this design to promote visualization as a learning strategy are the

- constructivist approach that encourages the student to articulate prior understanding, and to focus attention on key features of the prior mental model at the molecular level, *before* seeing the animations
- opportunity to discuss ideas and difficulties with peers
- practice, application, and explicit assessment of the visualization skills developed

The learning outcomes are to assist students in

- constructing scientifically acceptable mental models of substances and reactions at the molecular level
- relating these models to the laboratory and symbolic levels in chemistry
- applying their models to new substances and reactions
- using their models to understand new chemistry concepts that require a molecular-level perspective
- addressing common misconceptions identified in the research literature
- improving their confidence in explaining phenomena at the molecular level
- enhancing their enjoyment of chemistry by providing ways for them to use their imagination to explain phenomena, instead of just rote-learning terms and solving problems algorithmically.

How to Incorporate Multimedia in Your Chemistry Teaching: An Example of the *VisChem Learning Design*

The learning design is an attempt to make each stage of the multimedia information processing model—receiving, selecting, processing and encoding—as efficient as possible.

In this application of the design, we assume that students have had previous experience with visualizing simple pure substances—ionic compounds and water—and are familiar with conventions for representing molecules and ions. The specific learning outcome is for the student to visualize ionic solutions in terms of moving *hydrated* ions that occasionally form transient ion pairs. The most common misconceptions are that the ions do not interact with the solvent and that they are clustered together in "ionic formula units" (Figures 4 and 5). This is a problem if students are to understand other concepts, such as colligative properties.

Figure 4. Student's initial representation of an aqueous solution of barium chloride (Dalton, 2003).

Figure 5. Same student's later representation of an aqueous solution of barium chloride (Dalton, 2003).

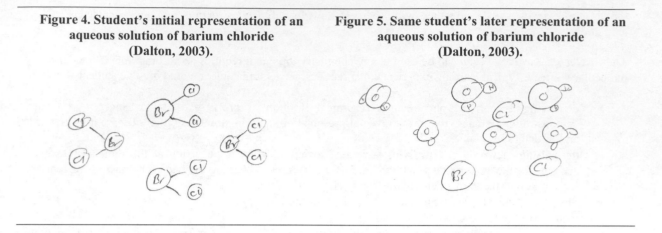

Suppose we start with a simple but interesting, phenomenon.

1.   Observing a phenomenon

In the first step, students write observations for a laboratory-level chemical phenomenon such as a physical property of a substance (e.g., a metal conducts electricity) or a reaction between substances (e.g., precipitation of an ionic compound). You can present this phenomenon as a live demonstration or with video (digital or analog), and you must ensure that *all relevant observations* are contributed by students.

For example, solid hydrated copper(II) sulfate and aqueous copper(II) sulfate solution are both light blue, but solid anhydrous copper(II) sulfate is white (Figure 6). What is/are the chemical species responsible for the blue color?

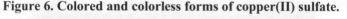

Figure 6. Colored and colorless forms of copper(II) sulfate.

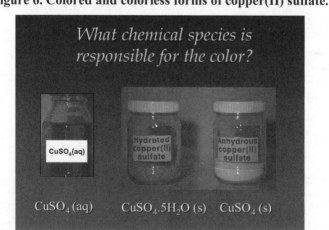

At this point, the instructor should allow students to think about and discuss this observation before rushing in with an explanation. An immediate but incorrect suggestion might be that, since all bottles contain copper(II) ions and sulfate ions, and only the blue forms contain water, perhaps water alone is responsible for the blue color. A moment's thought tells the student that this volume of pure water is colorless, so the answer must be more interesting!

The aim of this step in the design is to capture attention with an engaging context and then generate a "need to know." In terms of the multimedia information-processing model, the attention centers in the brain are being activated to pay attention to relevant aspects of visual and verbal information from the eyes and ears.

2.   Describing and drawing a molecular-level representation

In this step, students attempt to explain their observations by drawing *labeled molecular-level representations* of the substance or reaction, and also by describing their ideas in words. You need to develop the "drawing literacy" of your students by discussing conventions (e.g., using relative sizes of atoms and ions, using space-filling or ball-and-stick models, using keys, etc.), and pointing out that they will have to do such drawings as part of formal assessment. In this way, you are signaling that this is a skill worth developing.

With respect to the blue color, perhaps there is an interaction between the ions and the water molecules in the blue solid and solution. At this point, ask your students to represent their mental models of the chemical samples in all three bottles to their peers. This should be done in *both* words *and* diagrams to cater to students with a preference for expressing their ideas verbally or visually.

An alternative to drawing a representation of copper(II) sulfate solution is to use the Molecular Construction Tool (a free, downloadable program from the *VisChem Learning Design* website—Tasker, 2003). The advantage of the tool is the progressive feedback available at any stage of the construction process.

Figure 7. A sample screen from the *VisChem Molecular-Level Construction Tool* (Tasker, 2002) showing feedback on seven key features. The ion ratio is incorrect, and too many water molecules are not oriented for optimum H-bonding.

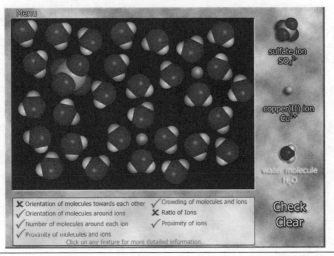

In terms of the multimedia information-processing model, the aim of this step in the design is to recall prior knowledge that will help the students focus on the key features of their representations.

3.   Discussing with peers

Following your advice to identify *key features* that explain the observations, students receive initial feedback on their representations by discussion with a peer (or from the Construction Tool). You should not identify correct or incorrect key features at this stage.

The feedback in the Tool is not designed to replace this discussion, but to focus attention on the seven key features of the representations that relate to crowding, proximity of molecules and ions, and ion hydration.

At this point, student attention will have hopefully been drawn to the key features, priming the perception filter (Figure 3) for selecting relevant verbal and visual information from the molecular-level animations that follow.

4.   Viewing animations

Animations can depict the dynamic molecular world more effectively than static pictures and words because students are spared the cognitive load of having to "mentally animate" the content. However, animations are *only effective if they are presented in a way that takes account of the limitations and processing constraints of the working memory.*

In this example, two animations depicting copper(II) nitrate solution are presented (see Figures 8 and 9). One animation shows all the water molecules and hydrated Cu^{2+} and NO_3^- ions, the other shows only the hydrated ions. Both animations are "busy," and, without prior experience with similar animations, the cognitive load on the working memory would be too high. However, by this point, the students' attention has been focused on particular key features, so they should *perceive* the hydrated ions.

The animations can be downloaded from the *VisChem Learning Design* website (Tasker, 2003), and they should be made available to students.

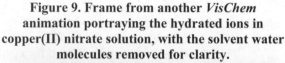

Figure 8. Frame from the *VisChem* animation portraying the hydrated ions in copper(II) nitrate solution.   **Figure 9. Frame from another *VisChem* animation portraying the hydrated ions in copper(II) nitrate solution, with the solvent water molecules removed for clarity.**

If there is enough time, each animation should be presented three times:

- First, without commentary, with students encouraged to look for the key features they identified in their own representations.
- Second, in "chunks" to reduce the load on working memory, each with narration by the lecturer drawing attention to the important key features, and with responses to questions from students.
- Third, in its entirety, with repeated, simultaneous narration.

5.   Reflecting on any differences with prior conceptions

In this step, students reflect on differences between key features in the animations and in their own representations, *amending their drawings accordingly, if necessary.*

Student drawings and descriptions of their conceptions of structures and processes at the molecular level often reveal misconceptions not detectable in conventional equation-writing questions. This activity in the learning design provides the opportunity for students to identify these misconceptions in their own representations, or those of their peers. *Experience shows that this is more effective than having the lecturer simply list common misconceptions.*

6.   Relating

In this step, you should encourage student discussion to link the key features of the molecular-level animations to the other two thinking levels.

Laboratory Level

- Can we see a relationship between the blue color and hydration of the copper(II) ions? If so, are the copper(II) ions bonded to water molecules in solid hydrated copper(II) sulfate?

Symbolic Level

- What is the ratio of $Cu^{2+} : NO_3^- : H_2O$ in a 1 M copper(II) nitrate solution? This enables students to visualize the term "concentration" in terms of "crowding" and to give some meaning to "1 M" compared with the concentration of pure water (1000 g/L = 55.6 M). The answer is 1: 2 : 56.
- How many water molecules, on average, are there between the ions in a 1 M copper(II) nitrate solution? This requires students to consider about 56 water molecules in a cube with one hydrated Cu^{2+} ion and two hydrated NO_3^- ions embedded within it. The answer is about two or three water molecules.

In terms of the multimedia information processing model, we are trying to link their new insight from the animations to their prior knowledge. This approach has been used successfully by others (for example, see Greenbowe, 1994)

7.   Adapting: Assessing Your Students' Understanding of Chemistry Concepts

In order to extend the links within the long-term memory, students are then shown an analogous substance or reaction at the laboratory level, and then asked to draw a representation. This establishes whether the students can transfer their ideas to a new example.

We have found that if students are to take visualization as a learning strategy, it is essential you encourage them to practice with homework problems and assess their visualization skills in your formal assessment. In addition to questions that probe qualitative and quantitative understanding of concepts at the symbolic level, we need to design questions that require students to articulate their mental models of molecular-level structures and processes.

One reason student misconceptions at the molecular level are not detected at the college level is that questions rarely probe this level of understanding explicitly. An example of a question style that does probe understanding at this level is shown in Figure 10. The student is required to draw representations using a key and to write text descriptions together with the conventional symbolic-level representations.

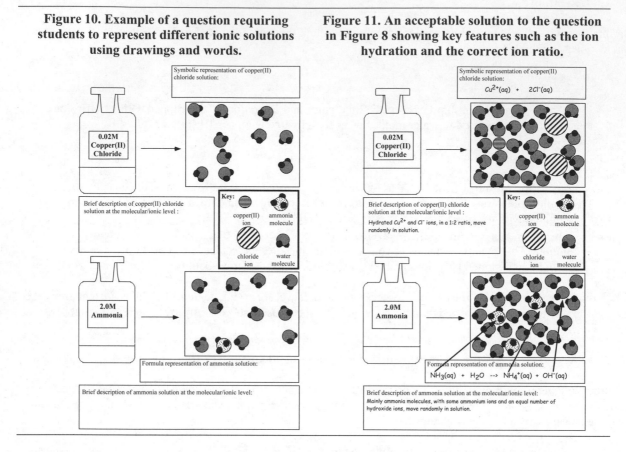

Figure 10. Example of a question requiring students to represent different ionic solutions using drawings and words.

Figure 11. An acceptable solution to the question in Figure 8 showing key features such as the ion hydration and the correct ion ratio.

Of course, these questions can also be answered without understanding, once "you know what the professor wants," like most other question styles. However, research with our students indicates that use of these questions has resulted in *more* students having scientifically-acceptable mental models (Dalton, 2003).

A good example of how you can probe deep understanding of difficult chemistry concepts by thinking at the molecular level is illustrated in Figure 12. Here, the question requires the student to know the difference between acid strength, acid concentration, and acidity (indicated by pH).

Figure 12. A question to examine whether a student understands the difference between acid strength, acid concentration, and acidity.

Compare the diagrams (**X**, **Y**, and **Z**) below, match each diagram to an acid solution (**A**, **B**, or **C**) in the following table, *and describe your reasoning*.

Solution	Acid	Concentration	K_a
A	trichloroacetic acid, CCl_3COOH	0.010 M	3.0×10^{-1}
B	chlorous acid, $HClO_2$	0.035 M	1.0×10^{-2}
C	benzoic acid, C_6H_5COOH	0.035 M	6.5×10^{-5}

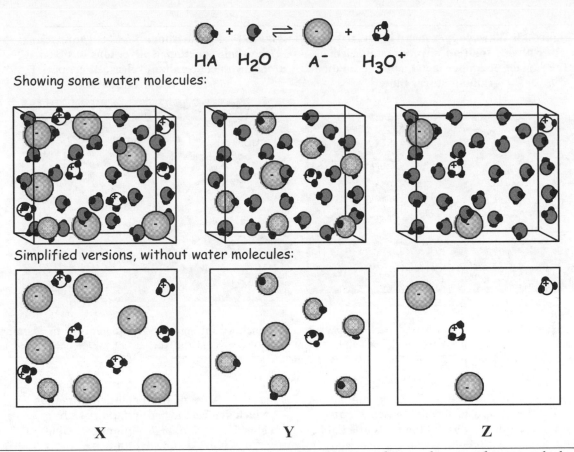

Showing some water molecules:

Simplified versions, without water molecules:

X   Y   Z

You might like to ask this question together with algorithmic questions that test the same ideas; see whether students can answer the latter without really understanding the concepts.

How can we improve a student's representational competence?

Students often have difficulties understanding and learning from visual representations. However, there is evidence that these difficulties can be overcome with appropriate scaffolding, particularly if students can generate and manipulate their own representations.

Representational competence is a term used by Kozma, Russell, and collaborators (1996, 1997) to describe a set of skills and practices that allow a person to use a variety of representations *reflectively*, to think and communicate about structures and processes at the molecular level. They have defined levels of this competence, and have developed the multirepresentational *ChemSense Knowledge Building Environment* software tools (*http://chemsense.org/*) to enable students to create their own representations.

What kinds of messages can be communicated in molecular-level animations?

Some of the *VisChem* animations described below can be viewed and downloaded from the learning design website (Tasker, 2002—*http://www.learningdesigns.uow.edu.au/exemplars/info/LD9/index.html*).

In contrast to most textbook illustrations, animations can show the dynamic, multiparticulate nature of chemical reactions. For example, the laboratory-level observation of silver crystals growing on the surface of copper metal shown in Figure 13 is hardly consistent with the misleading diagram, often found in textbooks, of one copper atom donating an electron to each of two silver ions. An animation (Figure 14) can show reduction of *many* silver ions on the copper surface, with concomitant release of *one half as many* copper(II) ions. This is a much better explanation for the 2:1 stoichiometric ratio in this reaction.

Figure 13. When copper metal is covered with silver nitrate solution, silver crystals form on the surface of copper metal, and the solution gradually turns blue.

Figure 14. Frame from a *VisChem* animation showing reduction of silver ions to silver atoms, with concomitant release of copper(II) ions, in a two to one ratio.

Animations of the molecular world can stimulate the imagination, bringing a new dimension to learning chemistry. What could it be like inside a bubble of boiling water, or at the surface of silver chloride as it precipitates, as depicted in Figures 15 and 16 respectively?

Figure 15. A frame of the *VisChem* animation, which attempts to visualize gaseous water molecules "pushing back" the walls of a bubble in boiling water.

Figure 16. A frame from another animation, which depicts the precipitation of silver chloride after mixing solutions of sodium chloride and silver nitrate.

Many molecular-level processes involve a competition between opposing forces. Examples include the competition for a proton on an iron(III)-bound water molecule by a solvent water molecule, and between lattice forces and ion-dipole interactions when sodium chloride dissolves in water, as shown in Figures 17 and 18, respectively.

Figure 17. Frame from a VisChem animation showing the "tug-of-war" for a proton on an iron(III) bound water molecule with a solvent molecule.

Figure 18. Frame from a VisChem animation showing the hydration of a sodium ion on the surface of sodium chloride despite strong attractive forces from the rest of the lattice.

How accurate do visualization models have to be?

In an earlier chapter, George Bodner quotes the Oxford English Dictionary definition of the term "model":

> *"A simplified or idealized description or conception of a particular system, situation, or process, often in mathematical terms, that is put forward as a basis for theoretical or empirical understanding, or for calculations, predictions, etc.; a conceptual or mental representation of something."*

Visualizing the invisible molecular world to generate useful mental models requires imagination and judgment. For example, the speed of atomic and molecular movements, as well as the uncertain non-Newtonian nature of electrons in atoms, requires substantial artistic license to enable the chemical structures and processes at this level to be communicated effectively. For this reason, students need to be constantly reminded that these animations are only models of reality. Indeed, progression from a novice to an expert in any science field can be imagined as simply a continual refinement of the scientific models in that field.

Great care has been taken in the *VisChem* representations of molecular structures and processes since research by Ben–Zvi et al. (1987) has indicated that misconceptions can be generated easily, and perpetuated, with poorly drawn images. The question is always *Will the artistic license result in "mental scarring"?*—that is, *Will it form restrictive misconceptions that will constrain future refinement of one's mental model?* The model of electrons in orbits around the nucleus is, arguably, an example of a mentally-scarring image that has been shown to be resistant to rejection in order to accommodate a much more useful quantum mechanical model.

Despite the care taken, small group interviews have revealed the potential for new misconceptions to be generated by some students from *VisChem* animations (Dalton, 2003). After viewing the animation in Figure 16, two students were curious about the reasons why the water molecules appeared to be "carrying" the silver chloride ion clusters towards the growing silver chloride crystal. This unintentional impression was communicated by not showing a sufficient exchange of hydrating water molecules around the ion cluster as it migrated towards the lattice.

Unfortunately, there is a wide range of quality in molecular animations available on the Web and as electronic supplements to textbooks. Compare the animation frame in Figure 19 below with the example in Figure 18. Both animations attempt to portray NaCl dissolving in water. What messages are communicated by these images?

Figure 19. Frame from an animation depicting solid sodium chloride dissolving in water. This animation was taken from a CD supplement to a college-level textbook.

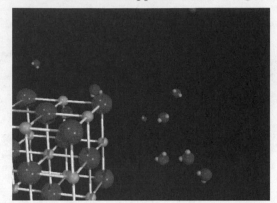

Each animation conveys implicit and explicit information. The animation in Figure 19 suggests that

- water molecules in the liquid state are widely separated with unhindered trajectories
- molecules and ions can cast shadows
- solid sodium chloride is composed of stationary ions separated from each other by stick bonds, with plenty of empty space. This image is reinforced in the animation when a water molecule passes between the ions in the structure (sic!) to hydrate an ion.

In contrast, the *VisChem* animation in Figure 18 portrays

- water molecules in the liquid state as much closer together (but not as close as reality)
- solid sodium chloride as composed of closely-packed ions that are *constantly vibrating*
- the competition between opposing, but almost-balanced, forces in the hydration process.

The frames in Figures 20 and 21 are taken from another example of a misleading animation, which is supposed to portray the reaction occurring when aqueous solutions of hydrochloric acid and sodium hydroxide are mixed. Figure 20 shows distinct HCl molecules in solution shortly before reaction with water molecules! Figure 21 shows NaOH "molecules" being added in a drop of solution.

Animations such as these reinforce the misconception that molecular formulas for strong electrolytes, such as "HCl(aq)," and so-called "molecular equations," such as

$$HCl \quad + \quad NaOH \quad \longrightarrow \quad H_2O \quad + \quad NaCl$$

actually describe the species present and the processes occurring at the molecular level. Little wonder students have trouble with understanding the nature of electrolytes and ion concentrations in solution!

Figure 20 is a frame from an animation depicting hydrochloric acid; it shows HCl molecules moving among water molecules shortly before reacting with them!   **Figure 21 is a later frame from the same animation, just before a drop of sodium hydroxide solution, containing NaOH ion pairs, is added from a tube above.**

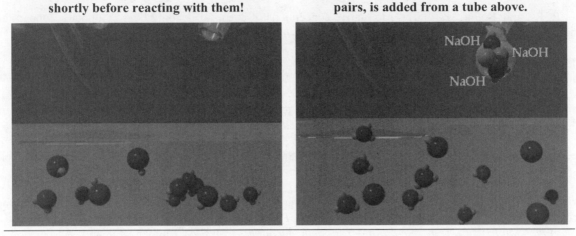

The above animation also illustrates the problem of mixing the laboratory and molecular levels in imagery. Could students develop the idea that a drop of water contains only a few ionic species? Could students develop the image of water composed of water molecules surrounded by some other watery matter indicated by the grey background?

We need to be very careful how we represent the molecular world to our students.

Conclusion

In the new millennium, most information is communicated using multimedia. The information-processing model based on the work of Mayer and Johnstone can guide us in developing effective learning designs using multimedia information.

Multimedia learning objects can communicate many key features about the molecular world effectively. However, we need to direct our students' attention to these features using constructivist learning designs that exploit our knowledge of how students learn.

We must assess our students' deep understanding of structures and processes at the molecular level. To do this effectively, we must encourage them to develop "representational competence" at this level.

There is a need to ensure that the same degree of rigor demanded from text-based information is applied to multimedia information. The latter is arguably more compelling and effective in facilitating conceptual change, for better or worse!

Suggested Reading

Tasker R. F., Bucat R., Sleet R., and Chia, W. (1996). The VisChem Project: Visualizing Chemistry with Multimedia. *Chemistry in Australia,* 63: 9, 395–397; and *Chemistry in New Zealand* 60: 5, 42–45.

Russell, J. W., Kozma, R. B., Jones, T., Wykoff, J., Marx, N., and Davis, J. (1997). Use of Simultaneous-Synchronized Macroscopic, Microscopic, and Symbolic Representations To Enhance the Teaching and Learning of Chemical Concepts. *Journal of Chemical Education.* 74(3), 330 - 334.

Sanger, M. J.; Phelps, A. J.; Fienhold, J. (2000) Using a Computer Animation to Improve Students' Conceptual Understanding of a Can-Crushing Demonstration. *Journal of Chemical Education.*, 77, 1517.

References

Ben–Zvi, R., Eylon, B.–S., and Silberstein, J. (1987). Students' visualization of a chemical reaction. *Education in Chemistry*. 24: 4, 117–120, 109.

Coll, R. K., and Taylor, N. (2001). Alternative conceptions of chemical bonding held by upper secondary and tertiary students. *Research in Science and Technological Education*. 19: 2,171–191.

Dalton, R. (2003). The Development of Students' Mental Models of Chemical Substances and Processes at the Molecular Level. *Ph.D. Thesis*.

Garnett, P. J., and Hackling, M. W. (1995). Students' alternative conceptions in chemistry: A review of research and implications for teaching and learning. *Studies in Science Education*. 25, 69–95.

Greenbowe, T. J. (1994). An Interactive Multimedia Software Program for Exploring Electrochemical Cells. *Journal of Chemical Education*. 71(7), 555–557.

Harper, B., O'Donoghue, J., Oliver, R., and Lockyer, L. (2001). New designs for Web-Based Learning environments . In C. Montgomerie, and J. Viteli (Eds.), *Proceedings of ED-MEDIA 2001, World Conference on Educational Multimedia, Hypermedia and Telecommunications*. 674–675. Tampere, Finland: Association for the Advancement of Computing in Education.

Hinton, M. E., and Nakhleh, M. B. (1999). Students' Microscopic, Macroscopic, and Symbolic Representations of Chemical Reactions. *The Chemical Educator*. 4(4), 1–29.

Johnstone, A. H. (1982). Macro and microchemistry. *School Science Review*. 64, 377–379.

Johnstone, A. H. (1991). Why is Science Difficult to Learn? Things are Seldom What They Seem. *Journal of Computer-Assisted Learning*. 7: 701–703.

Johnstone, A. H. (1997). "... And some fell on good ground". *University Chemistry Education*. 1: 1, 8–13.

Kleinman, R. W., Griffin, H. C., and Kerner, N. K. (1987). Images in chemistry. *Journal of Chemical Education*. 64: 9, 766–770.

Kozma, R. B., and Russell, J. (1997). Multimedia and Understanding: Expert and Novice Responses to Different Representations of Chemical Phenomena. *Journal of Research in Science Teaching*. 34: 9, 949– 968.

Kozma, R. B., Russell, J., Jones, T., Marx, N., and Davis, J. (1996). The Use of Multiple, Linked Representations to Facilitate Science Understanding. In S. Vosniadou, E. De Corte, R. Glaser, and H. Mandl. (Eds.) *International Perspectives on the Design of Technology-Supported Learning Environments* 41–60. NJ: Lawrence Erlbaum Associates.

Lijnse, P. L., Licht, P., Waarlo, A. J., and de Vos, W. (Eds.). (1990). *Relating Macroscopic Phenomena to Microscopic Particles*. Proceedings of Conference at Utrecht Centre for Science and Mathematics Education, University of Utrecht, and references therein.

Mayer, R. E. (1997). Multimedia learning: Are we asking the right questions? *Educational Psychologist*. 32, 1–19.

Moreno, R., and Mayer, R. E. (2000). A Learner-Centered Approach to Multimedia Explanations: Deriving Instructional Design Principles from Cognitive Theory. *Interactive Multimedia Electronic Journal of Computer-Enhanced Learning*. 2: 2.

Nakhleh, M. B. (1992). Why Some Students Don't Learn Chemistry. *Journal of Chemical Education*. 69, 191–196.

Nakhleh, M. B. (1993). Are Our Students Conceptual Thinkers or Algorithmic Problem Solvers? *Journal of Chemical Education.* 70(1), 52–55.

Nakhleh, M. B., and Mitchell, R. C. (1993). Conceptual Learning versus Problem Solving. *Journal of Chemical Education.* 70(3), 190–192.

Niaz, M. (1995). Relationship between Student Performance on Conceptual and Computational Problems of Chemical Equilibrium. *International Journal of Science Education.* 17(3), 343–355.

Nurrenbem, S. C., and Pickering, M. (1987). Concept Learning versus Problem Solving: Is There a Difference? *Journal of Chemical Education.* 64(6), 508–510.

Russell, J. W., Kozma, R. B., Jones, T., Wykoff, J., Marx, N., and Davis, J. (1997). Use of Simultaneous-Synchronized Macroscopic, Microscopic, and Symbolic Representations To Enhance the Teaching and Learning of Chemical Concepts. *Journal of Chemical Education.* 74(3), 330 – 334.

Taber, K. S. (1998). The sharing-out of nuclear attraction: or "I can't think about physics in chemistry." *International Journal of Science Education.* 20: 8, 1001–1014.

Tasker R. F., Bucat R., Sleet R., and Chia, W. (1996). The VisChem Project: Visualizing Chemistry with Multimedia. *Chemistry in Australia,* 63: 9, 395–397; and *Chemistry in New Zealand* 60: 5, 42–45. Available via W. H. Freeman textbook websites.

Tasker, R., Dalton, R., Sleet, R., Bucat, B., Chia, W., and Corrigan, D. (2002). Description of VisChem: Visualizing Chemical Structures and Reactions at the Molecular Level to Develop a Deep Understanding of Chemistry Concepts. *Available online at http://www.learningdesigns.uow.edu.au/exemplars/info/LD9/index.html.* Last accessed on November 25, 2003.

Tasker, R. F. (1992). Presenting a Chemistry Youth Lecture. *Chemistry in Australia.* 59: 3, 108–110.